Breeding of Horticultural Crops

Principles and Practices

Breeding of Horticultural Crops
Principles and Practices

Gopal Sharma

Oxford Book Company
Jaipur ▪ New Delhi

Published By

Oxford Book Company

Regd. Off. "Bony Residency", Gate No. 2
Opp. Tilak Public School, Vishweswariya Nagar,
Gopalpura Road, Jaipur - 302018, India
Ph: 0141 - 2761280, Telefax: 2761381
email - oxfordbookjpr@gmail.com
website - oxfordbookcompany.com

Delhi Office: 102, 1st Floor, Satyam House,
4327/3, Ansari Road, Darya Ganj,
New Delhi - 110002, India
Ph: 011-45652440

Distributed By
KEYA INTERNATIONAL PRESS LLP
E-mail: keyainternationalpressllp@gmail.com

Published in India, 2025

ISBN : 9789380179537

Typeset by Shivangi Computers, Jaipur

Printed and bound in India by Thomson Press Ltd.,
New Delhi

Contents

Preface

Plant breeding in the modern sense can be defined as purposeful genetic improvement. Yet genetic improvement of crop plants has an ancient tradition and in fact the greatest feat of plant breeding dates to Neolithic Revolution when our crop plants were domesticated starting about 10,000 years ago. Progress was achieved by selection of elite clones or lines of useful plants, mass growing of elite clones, followed by selection from naturally occurring seedlings. We recognize this technique as mass recurrent selection.

Plant breeding in the 20th century has accomplished a number of powerful achievements in agriculture. Perhaps the most important have been the increase in yield and adaptability of our major crop plants. At least half of the tremendous increases in yields of our major crops in the 21st century are attributed to genetic improvement. The development of photoperiod insensitive, short stemmed, fertilizer responsive rice and wheat—the Green Revolution of the late 1960s—is the high point of 21st century plant breeding.

The goal of this book is to address the principles and practices of horticultural plant breeding. In this respect it may be enlightening to distinguish between horticultural and agronomic crops because the case can be made that there is a fundamental difference between these crops that may affect the future course of genetic improvement. The difference between horticultural and agronomic crops is primarily one of usage. Horticultural crops are those that serve to fit the special food and esthetic needs of humans. They are crops that not only make life possible but make life worth living. In horticultural crops, quality is supreme.

The book has been designed for students, researchers in the field of agriculture, botany, horticulture, plant biotechnology, genetics and above all to a horticultural lover. With latest and updated information, the book has been compiled in a simple and easy language.

Dr. Radha Raman

1

Principles of Plant Breeding

Many people believe plant breeding is a post-Mendelian science. However, plant breeding in its rudimentary form started with farming about 12,000 years ago, in the form of selection of more suitable plants to man's needs throughout history. There are no records, but certainly one of the most important event was the domestication of crop and horticultural plants, such as wheat, barley, maize and apples. It is breathtaking to imagine that man's ancestors were able to understand that, through careful selection and replanting, they could change the nature of food plants in the directions they wished. There is some evidence that early-time biologists were crossing corn plants and producing hybrids well before Mendel's work. The onset of a scientific approach to plant breeding was perhaps the discovery of the cell and the role of sex reproduction in living organisms. It was Robert Hooke's discovery of cells in a piece of cork in 1665 that opened the door to many discoveries and inventions in biology and, especially, in genetics, although microorganisms had already been used for many years in the production of wine and bread. About 10 years later, Anton van Leeuwenhoek built a microscope with a 270 X amplification power, which allowed him to see microorganisms for the first time. The microscope opened the window to a new world, previously invisible to man. About 170 years later, Matthias Schleiden and Theodore Schwann presented the theory that all living organisms are constituted by cells.

It is not clear how much of intentional hybridisation, in order to provide new genotypes for selection, was performed before the scientific studies of sex role in plants started. On August 25, 1694, R.J. Camerarius published a paper in letter format entitled *"De sex plantarum epístola"*. Others may argue that plant breeding dates from the time of Thomas Fairchild's first artificial plant hybrid production (carnation x sweet william) in England in 1717, and Mather's cotton hybrid (squash x gourd) in America around the same time. In 1760, J.G. Kolreuter carried out the first scientific experiment in plant hybridisation; in 1823, Thomas A. Knight used hybridisation to improve crops and, by 1839, Carl F. von Gärtner had performed more than 1,300 different crosses. J.G. Kölreuter (1761-1766) published his plant breeding work in Germany describing over 100 experiments with artificial hybridisation, especially crosses, backcrosses, and F2 generations between tobacco species. During the 19th century, practical plant breeding started as business and, in 1819, Patrick Shireff in Scotland selected the first varieties of wheat using pure line selection. In France, Louis Vilmorin improved varieties of wheat and sugar beet using progeny tests. As knowledge evolved, new issues such as why offspring tend to resemble their parents called the attention of some scientists, but it was only at the end of the 19th century that the monk Gregor Mendel, working in Brno in the Czech Republic, unveiled the secrets of heredity.

Genetic knowledge to support plant breeding has advanced rapidly in the 20th century. By placing together a few more parts of the puzzle emerging from the findings of this new exciting science, scientists in the first half of the 1900's were able to conclude that something inside the cell was responsible for heredity. After the addition of dyes for visualisation of cells under the microscope, some structures that were stained in a distinct way were identified and named chromosomes. It was eventually learnt that these structures, the chromosomes, carried the genes that coded for the diversity of life. Although the genes determined human stature, eye color, hair color, and other traits, they could not be seen. However, their protein products could, and in 1941, George Beadle and Edward Tatum established a theory called "one gene, one enzyme", which answered a question that had persisted in the scientific community for many years: how genes provide instructions for building proteins. In 1944, Oswald Avery identified the DNA (deoxyribonucleic acid) as the gene's fundamental material. From this starting point, several research groups have devoted their time to study the DNA, and its chemical composition was quickly elucidated. The DNA is a molecule made of sugar, phosphate and four nitrogen bases i.e. adenine, cytosine, guanine and thymine (identified by the initials A, C, G, and T). Later, scientists realised that the four nitrogen bases, also known as nucleotides, were the alphabet of the genetic code.

The year of 1953 became a landmark for genetics with the discovery of the double-helix DNA structure by two scientists working at Cambridge University in England: James Watson and Francis Crick, an American and a British scientist, respectively. Their work revolutionised genetics and accelerated the discoveries of the fine DNA structure. The DNA capability to code for all the biological processes of living organisms relies in the genetic alphabet (A, C, G and T) which is written in the chromosomes. Genes differ in size (number of letters or nucleotides) and sequence (order of the nucleotides). For instance, a gene can have the following sequence: ATGCCGTTAGACTGAAA. However, one question remained: how the sequence of letter translated into proteins and traits.

Cracking the genetic code challenged geneticists. How could only four nucleotides code for the 20 different amino acids that constitute the thousands of existent proteins found in living organisms? It was only in 1967 that Marshall Niremberg and Har Gobind Khorana deciphered the genetic code. They concluded that the DNA was translated by groups of three nucleotides (or bases), later called codons. As there are four nucleotides

(A, T, C and G), there would be 64 different ways to arrange them into different codons. Taking into consideration that there are only 20 amino acids, the 64 different codons were more than enough to code for each of them.

GENETIC VARIABILITY AND EVOLUTION

The great navigators of the 15th century always brought home seeds of crops that were an important source of food or fiber. This idea was further developed to the introduction of new plants types, which carried genes of interest. Plant explorers were sent to different parts of the world to look for variability, the raw material of plant breeding. One of the most important explorers was a Russian named Nikolai Ivanovich Vavilov, who traveled extensively and identified eight areas in the world as centers of diversity of several crop species. He distinguished them between primary centers, which corresponded to the maximum occurring diversity suggesting the place where the crop was first domesticated, and secondary centers, which developed from plant types migrating from the primary center.

Vavilov considered these centers of diversity as centers of origin. He also demonstrated the existence of parallelism in variability among related species (the law of homologous series in heritable variation). The relevance of this law is that plant explorers could, to a certain extent, predict the forms not yet described, similar to what molecular geneticists are doing today based on sinteny between related species. Presently, many germplasm banks around the world collect and protect the available genetic variability, the raw material of evolution and commercial breeding to preserve the future of new generations.

Theory on Natural Selection and Evolution

Darwin's theory was that plants and animals vary in their traits, and that the variation is often heritable. Bestadapted individuals would survive the natural selection and would leave more descendents to the following generation, and since selection continues over generations, evolution was likely to take place. Because the mechanism of heredity was not known in that time, Darwin's ideas were not well accepted; and in the early decades of the 20th century, with the discovery of Mendel's work along with Johannsen's work, many biologists thought the Darwinism was dead. In the 1930s, the work of Fisher, Wright, Haldane, Dobzhansky, Huxley, Mayr, Simpson and Stebbins in different fields of mathematics and population genetic brought about an integration of the Mendelian genetics and Darwin's evolutionary theory. These works demonstrated that the Darwinian mechanism - natural selection- in Mendelian populations was enough to explain the evolutionary process observed in nature. The full depth and importance of Darwin's work on the adaptation and evolution of animals and plants was then clearly understood.

Cytogenetics

Cytogenetics aided not only the development of plant genetics but also the understanding of species relationships. Along with polyploidisation, cytogenetics played an important role in plant breeding efforts after the Second World War. The integration of cytogenetics to applied plant breeding was first done by E.R. Sears, who worked with wheat and by C.M. Rick, who worked with tomatoes. They demonstrated the possibility of transferring genes and chromosomes from alien to crop species, contributing to the use of cytogenetics in the evolutionary study of plant species. Cytogeneticists introduced a new genetic variation from many relatives in species of agronomic interest. This has been especially important in disease resistance breeding. While introducing race specific resistance genes from alien sources, breeders develop varieties with a more durable resistance. The best kwon introgression of alien germplasm accomplished by cytogenetics/plant breeding was the short arm of chromosome 1R from rye, which was substituted for chromosome 1B of wheat. The introduction of the 1BL-1RS translocation in wheat has been traced back to the work done in the 1920′ in Germany. The rye segment carried an array of useful disease resistance genes for leaf rust, yellow rust, stem rust and for mildew resistance. In other crops such as corn, translocations and chromosome introgressions were also important.

Mutagenesis

The pioneering experiments by Muller showed that it is possible to alter the effect of genes. This new possibility was looked upon with great optimism. However, after a few decades of research skepticism resulted from the disappointing experiments in generating useful mutations. After World War II funds became available for peaceful uses of atomic energy and many researchers in both developed and developing countries embarked on the fashionable technology of using radiation and radioactive isotopes for inducing mutations, often not paying attention to the rather modest results obtained between 1930 and 1945. In the following 20 years research mutation facilities were built around the world funded by the United Nations Atomic Commission. The International Atomic Energy Agency affiliated with the United Nations had supported the development of mutant varieties. Mutation breeding has now a more restricted but better-defined place among the methods of applied genetics. In addition to the use of induced mutations in crop improvement, there is now a most valuable use in fundamental genetics and plant physiology.

PLANT BREEDING THEORY AND RELATED ISSUES

Mendel's Laws

In the middle of the 19th century, the monk Gregor J. Mendel, working in Brno, in the Czech Republic, studied the genetic ratios of segregation, the first step to understand the secrets of heredity. Mendel noticed that traits of the garden pea plant, such as the flower color, plant height, and seed shape, were consistently transferred from parents to offspring. An inquisitive mind caused Mendel to make hundreds of crosses with different pea plants over many years. In recording the segregation of traits from thousands of plants, he discovered that they segregated in predictable ratios, season after season, indicating that the units of heredity were passed from generation to generation. These discoveries were presented in forums in the mid 1800's and compiled in Mendel's "Experiments in Plant Hybridisation" published in 1865. These were key discoveries to the emerging sciences of heredity, genetics and ultimately to plant breeding. Notwithstanding, Mendel's work was largely ignored during his lifetime. In fact, his research was virtually lost for 30 years, until it was independently discovered by three scientists, providing geneticists and breeders with a solid basis for handling trait heritability in substitution to some vague mystical effects.

Pure Line Theory

The experiences of plant breeders played their part in developing breeding methods, but it remained for Johannsen to place the individual-plant method of selection on a firm scientific basis. In 1903 and 1906 he studied field beans, a good choice because it is self-pollinated and contains traits that can be easily measured. He hoped to control heredity by applying Galton's law of regression, i.e. that the progeny of parents above or below the average tend to revert to average type. By selecting extreme individuals every generation, Johannsen observed improvement only in the first generation. From the second generation onwards all observed variability was due to the environment and was not passed to descendents. Johannsen's findings had a large effect on plant breeding, as he was the first to separate genotype from phenotype. He defined as pure-lined any individual obtained by continuously selfing the progeny from a cross between homozygous individuals.

Hardy-Weinberg Equilibrium

In 1908, Hardy in England and Weinberg in Germany independently demonstrated that in a large random mating population both gene frequencies and genotypic frequencies remain constant from generation to generation in the absence of mutation, migration, and selection. Breeding of cross pollinated species has since then been strongly based on the principles of the Hardy-Weinberg law.

Genetic Improvement Procedures/Breeding Methods

The genetic improvement procedures or breeding methods that are commonly used today were developed even before Mendel's discovery of genetics. They included mass selection, pedigree and bulk breeding. Later, in the first part of the 20th century backcross and recurrent selection were put forward. Other methods, such as single seed descent, were adopted much later. Selection of the best ear of corn or wheat for planting the following season was a practice used by farmers since the end of 17th century. However, the first scientific wheat-breeding programme started in 1898, by the Swedish Seed Association at Svalof, under Nilsson-Ehle. He used to hybridise wheat varieties to increase the genetic variability. Because of the large number of crosses, generations and plants involved each year, he invented the bulk method.

Louis de Vilmorin working in the 19th century in France was the first one to understand the effects of plant selection. Vilmorin developed the concept of plant selection with progeny test, which became known as the "Principle of Vilmorin". With his method he was able to carefully select beets based on their size, shape, and sugar content. Although the origin of the pedigree method is unknown, its first description can be attributed to Newman, who reported that until 1891, mass selection with self-pollination was predominantly used to improve self-pollinating species. The term pedigree selection was used when single-plant selection was performed to isolate pure lines in heterogeneous landraces. The pedigree method is still widely used with both self- and cross-pollinated species during population inbreeding.

Harlan and Pope emphasised the value of backcrossing for genetic improvement of crop varieties, observing that it had been used for many years in animal breeding to fix traits. Since then, backcross breeding has been used to improve varieties for individual traits, becoming a common method in some breeding programmes. At the University of California a series of small grain varieties was developed by backcrossing. The Atlas barley varieties were developed in the 1920 and 1930 using this method. It has lost its popularity among plant breeders in the second half of the 20thcentury. With the event of biotechnology new traits are coming to the market through using backcrossing to transfer them into elite lines. Thus, it is expected that backcross breeding will continue to play an important role in plant improvement. Goulden was the first to propose the concept of the single seed descent method (SSD). It did not catch the attention of many plant breeders until 1966 when Brim elegantly described it. Since then, SSD has become one of the most popular breeding method for inbreeding segregating populations in self- and cross-pollinated species.

Corn breeding methods have evolved over time with open-pollinated varieties, double-cross hybrid varieties and single-cross hybrid varieties, each one being popular in turns. H.K. Hayes and R.J. Garber at Minnesota, in 1919, and E.M. East and D.F. Jones, in 1920, independently suggested the concept of recurrent selection. Richey suggested the convergent breeding to increase vigor by recurrent cycles of selection. However, it was Hull who coined the term recurrent selection, which was immediately adopted. In 1949, F.E. Comstock, H.F. Robinson and P.H. Harvey proposed the use of reciprocal recurrent selection. Recurrent selection has been currently used in both cross and self-pollinated species, although it was developed with corn, a cross-pollinated grass. H.K.

Ideotype Breeding

The idea that plant physiology could contribute to plant improvement by helping understand physiological events drove many research institutions to devote time, resources and personnel to this area. The first major emphasis in breeding for ideotype was reported by Donald, an Australian biologist. In the U.S., Adams at Michigan State University and Rasmusson at the University of Minnesota applied the ideotype concept in common bean and barley, respectively, in the development of new varieties. Ideotype breeding may not have delivered all that it promised at first, but there is no question of its importance in shaping the way breeders handle their programmes. With the idea of pursuing a model, plant breeders were compelled to organise their thoughts and goals. Ideotype breeding also prompted breeders to introgress exotic germplasm in the breeding populations, expanding the genetic diversity used in varietal development.

Gene for Gene Hypothesis

Postulated by H.H. Flor, it introduced the concept that both host and parasite genetics played a role in determining whether or not a resistance reaction would be observed. Flor put forward the concept that the expression of resistance by the host was dominant while conversely the expression of avirulence by the parasite was dominant. This concept was carried forward by others, namely Clayton Person, in order to advance a theory that there was a single gene in the host that interacted with a single gene in the parasite. This theory was put forward at about the same time that the "one gene one protein" idea was being discussed by biochemists. Flor used this basic understanding when he formulated the "Gene-for-gene concept". Flor's tenet pointed out that basic host - parasite interactions are regulated by the genetic components of both the host and the parasite. Though that concept has been challenged by some, it is useful in understanding plant diseases. Flor's concept is still valid today and explains many hostpathogen interactions. A lot of research has been done in plant pathology since Flor's work and it is believed that about 25% of the plant breeding efforts in many crops have been devoted to the development of disease resistant varieties.

High Yield and Stability Breeding

High yielding and stable cultivars are the main goal of any modern breeding programme. Many high yielding modern cultivars are not commercially successful due to poor stability, while many highly stable cultivars do not provide adequate economic return. Breeding high yielding and stable cultivars is a major challenge facing breeders. Yield is essentially a complex quantitative trait controlled by numerous genes of individual small effects that are highly influenced by the environment. Stability can be achieved through the introduction of genes expressing resistance to diseases and insects, drought tolerance, nutrient extraction efficiency and many other factors. Breeding for stability also prompted breeders to introgress exotic germplasm in the breeding populations, expanding the genetic diversity used in varietal development. Further, it may help to preserve the environment for the future generations by reducing the use of agronomic inputs and allowing organic agriculture to develop.

Heterosis/hybrid Vigor

Although many plant breeders in the 18th century observed hybrid vigor, a clear understanding of the effects of self-fertilisation in cross-pollinated crops and of the effects of crossing self-pollinated plants was obtained only as a result of later genetic research. In 1876, James Beal, at Michigan Agricultural College carried out the first controlled crosses of corn to increase yield. George H. Shull started inbreeding open-pollinated corn at the Station for Experimental Evaluation, Cold Spring Harbor, NY. East and Hayes pointed out the value of hybrid vigor both in evolution and in plant breeding. The original concept to commercially exploit heterosis was the production of single-cross hybrids, but the costs of seed production initially limited its usefulness. This limitation was overcome with Jones's suggestion that double-cross hybrids can be produced from two single-cross hybrids to reduce costs of seed production and, consequently, doublecross hybrids became rapidly accepted worldwide. Since 1960, single-cross hybrids have been replacing double-cross hybrids in many places in the World.

Statistical Principles and Experimental Designs

Ronald Fisher, considered the father of modern statistics, worked at the Rothamsted Agricultural Experiment Station as a biologist and made many contributions to both statistics and genetics. He introduced the concept of randomisation and analysis of variance procedures, now used throughout all biology. He also introduced the concept of likelihood and gave a new definition to statistics. Its purpose was the reduction of data and he identified three fundamental problems. The contributions Fisher made included the development of methods suitable for small samples, the discovery of the precise distributions of many sample statistics and the invention of analysis of variance. He introduced the term maximum likelihood and studied hypothesis testing. Statistics arose in synchrony with ideas about quantitative genetics and gave breeders ways to deal with measurements and scores of quantitative traits. Statistics was essential to the formulation of theories about quantitative genetics, and to practical application of those theories.

Quantitative Genetics

Because single genes explain only a limited number of traits that mattered to breeders, quantitative genetics came about to give rational ways to deal with inheritance of quantitative traits. The boom of quantitative genetics started around the 50's, when Jenkins' work about the diallel designs was developed. In the 60's, quantitative genetics was so mainstream that in just about all scientific meetings it would be main theme.

Quantitative genetics has helped to understand and account for G x E, adaptability and stability as well as in the interpretation of data collected in breeding experiments. One of the most important contributions of quantitative genetics to plant breeding was the partitioning of the expected mean square into the covariance between relatives. The paper by Cockerham published in the Crop Science journal was paramount in connecting statistics (sources of variation and variance components) to genetics (covariance among relatives). The following work of Falconer in Scotland, Mather and Jinks in England, Eberhardt and Comstock in the USA and Vencovsky in Brazil, among others, further developed this important branch of the genetic science, providing many handy tools for unraveling the intricate genetic control of quantitative traits.

Green Revolution

The green revolution involved the simultaneous development of new varieties of food plants and altered agricultural practices that greatly increased crop yields. It became a worldwide agricultural movement which began in Mexico in 1944, when the Rockefeller foundation and the Mexican government established a plant-breeding station with a goal of boosting grain yields. Norman Borlaug, a plant breeder from the University of Minnesota headed the project. His research lead to high-yielding wheat plants. Later, Borlaug won the Nobel Peace Prize for his work that reduced famine in Asia and other regions. Borlaug's wheat varieties were successful when grown in many areas of Asia and Africa. In India, wheat production increased four times in 20 years. The successes in Mexico with wheat led to the establishment of other breeding programmes emphasising short stature plants. In fact the green revolution subsequently occurred Worldwide in many crops including rice, sorghum, maize, cassava, and beans. Borlaug basis for the Green Revolution was the use of short "semi-dwarf" wheat germplasm in his breeding programme. The semi-dwarf genes increased the harvest index in the new varieties, making them yield significantly more that the conventional varieties.

Breeder's Intellectual Property

The interest of the industry in breeding self pollinated crops became more evident after the Plant Variety Protection Act (PVPA) was enacted in December 1970 and amended in 1994. This provided legal intellectual property rights protection to developers of new varieties of plants that are sexually reproduced (by seed) or are tuber-propagated. Bacteria and fungi are excluded. The PVPA is administered by the United States Department of Agriculture. A Certificate of Protection is awarded to an owner of a variety after an examination shows that it is new, distinct from other varieties, and genetically uniform and stable through successive generations.

The term of protection is 20 years for most crops and 25 years for trees, shrubs, and vines. The owner of a U.S. protected variety has exclusive rights to multiply and market the seed of that variety. Some scientists still believe that the breeder's intellectual property also had a negative impact in plant breeding, reducing the exchange of information and germplasm among breeders and companies. The breeder's intellectual property has had a major impact in plant breeding since 1990 and it is expected to have an even larger impact in the future, especially with the biotech traits introgressed in the new varieties. In Brazil, the intellectual property issue has also been dealt with since the Paris Convention in 1835. But it was only in 1994, when the Brazilian Congress ratified the "Agreement on Trade Related Aspects of Intellectual Property Rights" (TRIPS) of GATT that the issue was discussed for legislation purposes. In January 1995 and in April 1997, the Brazilian Biosafety legislation and the Variety Protection Law were respectively approved.

BIOTECHNOLOGY IN PLANT BREEDING

Discovery of DNA Structure

The Watson-Crick hypothesis provided chemical explanation and high accuracy to Mendelian genetics and led to the development of molecular biology, which allows breeders to know intimate details of genes and to manipulate them. At the moment, however, this science is at about the same stage as when Mendelian genetics was established but quantitative genetics had not been developed. Molecular biology gives too much detail and too much precision as far as breeders are concerned at this time. At this time geneticists/breeders are trying to develop something akin to quantitative genetics to deal with the masses of information they have accumulated on genes, gene products, gene regulation and pathways of action and interaction.

Biotechnology

Biotechnology is the application of biological research techniques to the development of products or processes, using biological systems, living organisms, or derivatives thereof. It can also be understood as a range of different molecular technologies such as gene manipulation and gene transfer, DNA taping and cloning of microorganisms, plants and animals. Under biotechnology is also tissue culture, genomics and bioinformatics.

Tissue Culture

One of the major breakthrough in tissue culture occurred when Murashige-Skoog developed the MS media in 1962 containing nutrition factors that allowed the *in vitro* growth of many tissue types. Tissue culture encompasses many techniques that can be applied to plant breeding, such as embryo rescue, anther culture, micropropagation, indexation, *in vitro* selection, and somaclonal variation. Anther culture is important especially in crosses between individuals from different species. In those cases, it is frequently necessary to rescue the embryo in the developing seed to obtain the interspecific hybrid.

Anther culture has been widely used in the breeding of wheat, barley, corn and many other species. It provides a fast mean to reach homozygosity after crossing two parents. While using conventional breeding methods such pedigree, bulk and SSD may take from 4 up to 7 or more years to reach homozygosity, with anther culture it is possible to substantially shorten this period. Some varieties have been developed using anther culture in different countries around the World. Tissue culture has also been important to provide means for the regeneration of whole individual plants in gene transformation systems such as biolistic or *Agrobacterium* mediated transformation. Therefore, tissue culture is expected to play an even more important role in cultivar development in the future.

Molecular Markers

Molecular markers are DNA fragments that may be used as a fingerprint in the identification or characterisation of individuals. These markers have become an increasingly helpful tool in genetic research, and in biotechnology applications. The basic premise behind molecular markers is that there is a natural genetic variation in individuals, and many genetic sequences are polymorphic, meaning that they differ between individuals. Molecular markers seek to exploit this variation to identify individuals, traits, or genes based on genetic differences. Two of the most common molecular markers used in plant breeding are: Restriction Fragment Length Polymorphism (RFLP) and Randomly Amplified Polymorphic DNA (RAPD). In recent years more and more molecular markers have become available. Some of them are STR (Short Tandem Repeats or Microsatellites), STS (Sequence Tagged Sites), Expressed Sequence Tags (EST), DNA Chips, SNP (Single Nucleotide Polymorphism or point mutations), AFLP (Amplified Fragment Length Polymorphism), and SCAR (Sequence characterised Amplified Regions). They have been used to establish a link between genes and traits, and to determine the location of genes in the genome and in marker assisted selection. With the detailed maps and probes for genes of known function, it has become possible to discover close linkages of a molecular marker with a trait of interest. This information has been used to move genes from one genetic background to another.

Gene Transformation

Genetic transformation may be regarded as a century event in genetics and in plant breeding, in the sense that creating transgenic organisms enables breeders to develop varieties never thought before (despite the utility of and the great excitement and/or fear about Bt maize/cotton and RR soybeans). Such single gene changes can be very nice additions, but single gene changes are not fundamental to making improved varieties. On the other hand, genetic transformation is a very powerful tool for learning about gene functions. Transgenic breeding creates "artefactual" variants *de novo* and transgress the barrier between species. It will help in the real task for breeders: to manipulate the regulation, interactions and pathways of operation of many genes at once in order to make a well-balance organism. Molecular plant breeding for the construction of organisms that make entirely new products such as pharmaceuticals or chemicals would not be possible without genetic transformation.

Genomics

In recent years, technological advances in molecular biology and bioinformatics have consolidated genomics, putting genetics into a new age. Genomics is a science that uses molecular characterisation, transcript profiling and cloning of whole genomes to understand the structure, function and evolution of genes as well as to answer fundamental biological questions. It is a brand new science that promises to revolutionise plant breeding and genetics itself. Probably, the main contribution of genomics to plant breeding will be the knowledge generated, which will allow breeders to accomplish their goals in a much faster and predictable manner. For example, genomics will reveal the underlying function of genes, placing plant models (ideotypes) within reach of plant breeders without the linkage drag. When genetics came about at the beginning of the 20th century, it shed light in the breeding work done by practical plant breeders. Genomics is shedding new light into plant breeding.

The Arabidopsis genome and rice genomes, which are models for dicots and monocots respectively, were sequenced in 2000 and April 2002. The sequencing of other crop genomes will come at a faster rate than previously thought. The knowledge of model genome sequences will allow scientists to reveal genetic principles putting the science of genetics into a new perspective. For example, the sequence of the *A. thaliana* genome has revealed the genes and principles of flowering. Scientists have then manipulated yellow poplar (*Populus trichopulus*), developing varieties of this tree that blossom when they are one year old, much sooner that other conventional varieties.

REFERENCES

Bhat, S.R. and Srinivasan. 2002 "Molecular and genetic analyses of transgenic plants: considerations and approaches". *Nature*. 349:726.

Crow, J.F. 1998. 90 "Years ago: the beginning of hybrid maize". *Genetics*. 148:923-928.

Harborne, J.B. 1978. "Plant tissue culture and its biotechnological application". *Phytochemistry*. 17:1691-1692.

McCann, D.A and Batterham, P. 1993. "Australian genetics: a brief history". *Genetica*. 90:81-114.

Olsson, G. 1986. *Svalof 1886-1986- research and results in plant breeding*. Svalof AB Press, Svalof.

Pauls K.P. 1995. "Plant biotechnology for crop improvement". *Biotechnology Advances*. 13:673-693.

Richey, F.D. 1927. "Convergent improvement in selfed alfalfa". *Am. Nat.* 61:430-449.

Shull, G.H. 1908. "The composition of a field of maize". *Am. Breeders' Assoc. Rep.* 4:296-301.

Tsitrone, A; Rousset, F and David, P. 2001. "Heterosis, marker mutational processes and population inbreeding history". *Genetics*. 159:1845-1859.

2

Conceptual Framework for Plant Breeding

Population projections by the United Nations indicate that the world's population will increase to approximately 8.2 billion persons by 2030. Other estimates vary between 7.6 and 8.8 billion depending on the utilized assumptions, and the numbers have been continuously revised- mostly downward- in the last few decades to reflect changes in birth and mortality rates and the effects of natural and man-made disasters. However, all of these projections agree on two basic issues: first, the world population will continue to grow, and second, this growth will occur mainly in the less and least developing countries, which are ill-equipped to supply increasing food demands from growing populations.

Some authors have suggested that the problem of world hunger is one of unequal distribution rather than quantity, as there is currently sufficient food to feed the world's population. However, even if the distribution problem could be solved today, larger food amounts will still be needed for the growing population of tomorrow. In the context of the global agricultural food system, the challenge is not only the need to feed an increasing population, but to do so within productivity constraints that are not always addressed by policy makers and scientists. Such productivity constraints may include limitations on the amount of arable land and non-renewable resources (particularly water), increases in urban populations, decreases in the number of people engaged in agriculture, the effects of climate change in agriculture, and increased competition for land due to urbanization or other land uses.

To date, productivity constraints have been addressed by technological innovations brought about through investments in research and development. Agricultural innovations have contributed to poverty alleviation efforts in the past by attempting to reduce vulnerability and in many cases by enhancing and/or increasing a given community's asset base and/or productivity. These innovations have had different degrees of success in influencing the policies, institutions and processes in rural communities, and the development of alternative or better strategies to support livelihood improvement.

One such agricultural innovation is the use of improved crop varieties, landraces, hybrids, and other plant genetic materials. Collectively, genetic materials conserved and used for breeding have been referred to as 'plant genetic resources for food and agriculture,' or PGRFA. These unique and diverse resources have been the backbone of crop improvement for centuries, ever since crop domestication began. New varieties are continuously derived from PGRFA and are essential for agricultural improvement, just as they were during the Green Revolution, when PGRFA helped raise production levels, provide greater food security, and increase the incomes of numerous farmers in the developing world. Newer biotechnology techniques also rely on PGRFA to ensure transfer of valuable traits for the benefit of poor farmers.

Farmers around the world can benefit from crops that have higher (maintained) yield potential, increased productivity, and cutting-edge resistance to biotic and abiotic stresses. These improvements, however, require public and private investment in plant breeding and biotechnology. Applications from modern biotechnology have the potential to be useful in addressing specific needs of the poor, but cannot realize their potential without investments in plant breeding research. Investments in pro-poor plant crop improvement research (including those that use biotechnology) may be hampered by a shift towards privatization and the inability of seed systems to deliver innovations to farmers. Conventional plant breeding has been shown to account for 50-60% of yield increases, but plant breeders around the world continue to face challenges similar to those of the past, including the need to increase productivity, improve environmental quality, and provide new crop options for farmers.

GOAL AND OBJECTIVES

Despite the recognized importance of reliable information on crop improvement, minimal data are available for public and private plant breeding in developing countries. To address this gap, the FAO launched a research project with several national and international partners in 2002. This research project included the design and implementation of a survey of all relevant organizations in multiple countries around the world. The FAO survey was designed to answer two main questions: (1) how are plant breeding programs directing financial, human, and institutional resources towards innovations derived from PGRFA that will help address food and production needs; and (2) based on the information analyzed, what policy changes can be recommended? To answer these questions, the survey examined and contrasted five critical factors over a 16-year timeline. The critical factors included:

1. institutes with mandates for plant breeding and biotechnology,
2. the trends and focus of each country's plant breeding research,
3. full-time equivalent (FTE) scientists working in plant breeding/biotechnology and their associated research budgets,

4. PGRFA used for cultivar/variety development, and
5. capacities and approaches for crop biotechnology and breeding.

After data collection was completed, an in-depth, standardized analysis of the trends and conditions for plant breeding in the public and private sector was undertaken by the FAO in collaboration with IFPRI and other Consultative Group for International Agricultural Research (CGIAR) centers. Here, we use the results of the in-depth analysis made by IFPRI on the FAO survey data collected for Cameroon, Kenya, the Philippines, and Venezuela. The data collected in the surveys measured the plant breeding and biotechnology capacity of each country, and their ability to deliver innovations based on PGRFA.

Although not an explicit objective of the FAO survey, the study team frame their discussion within the broader context of an innovation capacity and systems framework. In addition, we outline a conceptual framework for plant breeding and biotechnology innovations based on studies done by Furman et al. and others. The innovative systems approach allows a broader and inclusive conceptual foundation, from which we derive potential policy instruments and approaches for the four countries included in the study. The innovation framework described here allows us to propose strategies for the four studied countries (and similar countries) and supports status mapping based on a country's current plant breeding and biotechnology capacity and policy status relative to the overall strengths, limitations and opportunities of their national innovative capacity. In addition, the innovation framework allows us to identify explicit determinants of innovative capacity that may be used to pinpoint the gaps, limitations and/or strengths of the overall innovation system containing the components of plant breeding and biotechnology. Within this framework, it is possible to delineate a set of relevant policy interventions for internal discussion.

CONCEPTUAL FRAMEWORK

The design and implementation of national policies aimed at addressing constraints can be significantly improved if these policies are drawn from a broad, inclusive and flexible conceptual framework. Such a conceptual framework will help develop better policy interventions by identifying the relative strength of intervening factors (the "determinants of innovation") and by supporting strategic interventions based on the current and expected capacity to implement in-country research. Countries may use a conceptual framework to identify areas where gaps, limitations, and threats are rendering the innovative system inefficient, ineffective or unresponsive. This approach, in conjunction with a robust priority-setting process, development of policy matrixes, and other strategic approaches for technology assessment, will provide additional support for policy development. However, proper policy development requires that the chosen conceptual framework consider all stakeholders, linkages and institutions within the innovation and technology transfer system. This is the case with plant breeding, biotechnology, and PGRFA, which converge to produce crop bio-innovations for farmers.

The FPS conceptual framework considers two distinct levels of innovation and their linkages. At an aggregate (national) level, the FPS model considers those determinants of innovation common to all innovative activities, which are included under the group titled "Common Innovation Infrastructure." The Common Innovation Infrastructure is the foundation of a nation's ability to support innovative activities, and in some cases even enable new ones. While it may be possible for specific groups or firms in a country to create innovation in the absence of a common national innovation infrastructure, the long-term national capacity to create innovations across a broader spectrum of disciplines is significantly hindered by the absence of a common innovation infrastructure. The minimum common innovation infrastructure needed to support group-specific (i.e. plant breeding and biotechnology) innovation is likely to vary from country to country, but it is critical to first view the system as a whole, and then examine the factors likely to affect innovation at the group-specific level.

Many innovations arise from specific groups conducting research. These groups, called "clusters" by Furman, Porter, and Stern (and other authors), are the basic unit of innovative capacity, and may consist of a group of researchers, institutes, firms or consortia of research teams. Each individual cluster is connected to other related clusters, and they interact to support innovative capacity. Furthermore, each individual cluster (or group of clusters) is subject to a set of factor (input) and output demand conditions, as well as a firm-specific context of strategy and rivalry. At the same time, each cluster is closely tied to activities among related and supporting industries. This is a very dynamic process, where opportunities and ideas arise through the strength of the innovative system. The main objective of the FPS framework is to describe national innovative capacity by simultaneously examining the common infrastructure, linkages, and multiple clusters pertaining to innovation within a nation.

FAO PLANT BREEDING SURVEY DEVELOPMENT

Frey conducted a short, comprehensive survey of all plant breeding research institutes in the United States of America to identify and quantify all human and financial resources in the area. Thereafter, FAO identified the need to extend this seminal survey to national plant breeding programmes in developing countries. The FAO survey was designed by a team of experts that met at FAO in 2003; this collaboration yielded an FAO project aimed at complete national surveying of plant breeding research and resources in developing countries. As of October 2007, the FAO surveys had been completed in 50 countries and were in-progress in an additional 23 countries. Well recognised in-country plant breeding experts were recruited to conduct the survey in their respective countries. These in-country consultants were asked to survey all public and private sector organisations involved with plant germplasm conservation, enhancement and use in their respective countries.

FAO Survey Data Verification

The original survey was designed to collect information up to 2001. At IFPRI's request, some stakeholders updated the information to 2004, as important changes had occurred over the intervening years. In addition, IFPRI and other participants spoke at an FAO plant breeding meeting, requesting collection of additional data on the age distribution of human resource capacity, which had been identified as a critical issue. FAO was able to request this information from countries that were still in the process of completing their surveys, but budgetary constraints precluded data collection from all countries on this parameter. The survey requested information on average overall budgets for plant breeding and biotechnology, and asked for a breakdown of percent allocation of resources. The rationale behind this approach, following that of Frey, was to obtain an estimate from responsible individuals within the organisation, and to override their natural hesitancy to provide information that may be viewed as confidential (by private firms) or may be hard to recall (as is sometimes the case for the public sector). Each national consultant collected data across five general categories:

1. institutional information,
2. human and financial resources dedicated to plant breeding and biotechnology,
3. effort allocation by crop,
4. sources of germplasm, and
5. output in terms of varieties released.

Since the Rome meeting in 2005, IFPRI has contacted four national experts from the Rome meeting to clarify inconsistencies and gaps in the country-level data. However, at present there are still questions regarding some of the responses in each country. In particular, the data on budgets and financial resources may be somewhat questionable as identified a variation in terms how to measure financial resources and what to include in such estimation. From this standpoint, data on financial resources has to be viewed with some caution.

Innovative Capacity and the Mapping of Countries

In Table 1, the general framework proposed by Fuglie and Pray, which seeks to explain innovation and technology transfer of biotechnology innovations in terms of explanatory factors. The Fuglie and Pray approach explains biotechnology innovation using four distinct factors or "determinants" of innovative capacity: crop improvement, biotechnology capacities, field trials, and market size. Each of these determinants is in turn described by a distinct set of innovative influencing policies. As can be seen in Table 1, use these indicators to cluster countries based on their functional capacity, and derive generic policy situation categories and tools for further analysis. The approach taken in this chapter is similar to that outlined by Trigo, who uses a set of variables to rank countries in terms of their current potential to implement biotechnology and/or plant breeding. With the variables included in his framework, Trigo classifies countries according to three policy situations (or stages) ranked here from more to less advanced in capacity:

1. Building capacities to develop biotechnology based innovations.
2. Improving the efficiency and products of agricultural research through the use of biotechnological tools (classified under large, medium and small markets).
3. Setting the stage for using biotechnology products.

Table 1: Innovative outputs for developing economies in the FAO survey

Innovation Output	Indicator	Unit	Philippines	Kenya	Venezuela	Cameroon
Improved varieties release	Total number of improved varieties produced	Number	138(215)	153	192	100
	Average of total number of varieties produced	Number	43	33	48	25
	Average growth rate of the number of varieties produced by public/private sectors across years and crops	%	78	126	13	–36
Biotech products	Biotechnology approved for confined/ small-scale field trials to 2006	Number		3	0	0
	Biotechnology products approved for food/feed consumption and number of crops to 2006	Number/ number	37/7	0/0	0/0	0/0

COUNTRY DATA ON AGRICULTURE

The four countries examined in this report have very distinctive characteristics, not only because they are located on three different continents (Africa, Asia, and Latin America), but also because their main socioeconomic indicators are quite different (See Table 2). Cameroon is the smallest of the studied economies, both in terms of its Gross Domestic Product (GDP) and population, although Kenya has a smaller GDP per capita than Cameroon. In contrast, Venezuela has the largest economy in the group, followed by the Philippines. Cameroon is mainly an agricultural-based economy, with the agricultural sector contributing roughly 44% of the GDP and at least 20% of all exports in 2004. In contrast, agriculture is one of the smallest sectors in Venezuela, representing just 4.5% of the GDP and contributing less than 1% to total national exports in 2003. The Philippines and Kenya fall between these extremes, with their Agricultural GDPs (Ag GDPs) varying from 19%-30% and 15%-22% for Kenya and Philippines, of all economic activity for the averages of the three four-year periods from 1991 to 2003.Cameroon is the only studied country in which the Ag GDP has increased over the past 10 years, with agriculture as a percentage of GDP growing from 28% in 19936 to 44.2% in 2004. Although Cameroon and Venezuela are both oil-rich countries with high fuel exports (50% and 82% of total exports, respectively), they differ substantially in terms of economic performance. Cameroon has a vigorous agricultural sector and sustained a relatively steady economic growth at an average annual rate of 4.3% during 1995-2003, whereas the Venezuelan economy declined at an average rate of 3.7% during that same period. The Philippines experienced a growth similar to that of Cameroon, while Kenya registered a growth rate of only 1% during the 1999-2003 periods. In terms of land, the Philippines represent the smallest area (29 million hectares) and Venezuela the largest (88 million hectares). However, Venezuela has only 2.6 million hectares of arable land, compared to 4.5-6 million hectares in the other studied countries.

Table 2: Basic economic, production, and population data for study countries

Indicator	Country	Average 1991-1994	Average 1995-1998	Average 1999-2003
Agriculture, value added (% of GDP)	Cameroon	30	42	44
	Kenya	30	29	19
	Philippines	22	20	15
	Venezuela	5	5	4
Agriculture, value added (constant 2000 US$)	Cameroon	2,420,666,784	3,064,765,817	4,032,175,922
	Kenya	1,593,777,536	1,739,067,186	1,808,417,223
	Philippines	10,337,120,387	11,045,273,223	12,399,414,375
	Venezuela	4,524,134,441	4,681,181,172	4,790,506,939
GDP growth (annual %)	Cameroon	–3.2	4.6	4.6
	Kenya	0.9	3.1	1.0
	Philippines	1.6	3.8	4.3
	Venezuela	3.4	2.6	–3.7
GDP per capita (constant 2000 US$)	Cameroon	569	545	604
	Kenya	360	361	346
	Philippines	886	946	1003
	Venezuela	5429	5333	4691
Land area (hectares)	Cameroon	46,540,000	46,540,000	46,540,000
	Kenya	56,914,000	56,914,000	56,914,000
	Philippines	29,817,000	29,817,000	29,817,000
	Venezuela	88,205,000	88,205,000	88,205,000
Land use, arable land (hectares)	Cameroon	5,957,500	5,960,000	5,960,000
	Kenya	4,225,000	4,350,000	4,550,000
	Philippines	5,496,750	5,512,500	5,662,500
	Venezuela	2,657,750	2,587,750	2,556,500
Population, total	Cameroon	12,538,758	13,931,825	15,440,691
	Kenya	25,014,750	27,705,500	30,700,838
	Philippines	64,608,033	70,737,140	78,253,712
	Venezuela	20,890,000	22,729,250	24,767,400

Cameroon

Institutional Information

In Cameroon, research activities in plant breeding and biotechnology are carried out by a handful of public institutes, particularly the national institute of agricultural research for development (Institut de Recherche Agricole pour le Développement, IRAD). Others relevant institutes include two universities (University of Yaoundé and University of Dschang—FARA, Agronomy and Agricultural Science Faculty), and the Banana and Plantain African Research Center. Table 3 shows details on all four institutions surveyed and their years of experience with biotechnology and plant breeding.

Table 3: Participating countries, numbers of institutes, and experience with plant breeding and biotechnology by country

Country institutions	Contact expert institution	Surveyed experience	Type of experience	Plant breeding	Biotechnology
		Number		**years**	**years**
Cameroon	Dr. Joseph Kengue	5	Public	15-50	14-18
Kenya	Dr. Josephine Songa	10	Public	7-85	0-22
			Private	1-27	0
Philippines	Dr. Desiree Hautea	5	Public & Private	18-80	5-26
Venezuela	Prof. Orangel Borges	3	Public & Private	14-53	0-13

Human/financial resources dedicated to plant breeding and biotechnology. Table 4 presents the distribution of scientists by degree and whether they conducted research in plant breeding or biotechnology during the study period. The total number of scientists involved in plant breeding research increased from 27 in 1995 to 38 in 2001. The increase in total number of scientists involved with plant breeding was the result of increases in the number of scientists' at all educational levels.

Table 4: Distribution of human resources by degree, area of expertise, and country

		Plant Breeding				Biotechnology			
		B. S.	M. S.	Ph. D.	Total	B. S.	M. S.	Ph. D.	Total
Cameroon	1985	10	9	8	27	—	—	3	3
	1990	27	10	13	50	4	4	9	17
	1995	16	11	12	39	8	4	8	20
	2001	13	12	13	38	7	5	6	18
Kenya	1985	20	16 4	39	0	0	5.3	4.3	
	1990	8	27	8	42	0	0	5.7	5.7
	1995	8	22	20	50	0	0	5.3	5.3
	2001	7	25	20	52	1	0	5.3	6.3
Philippines	1985	56	33	9	98	8	3	1	12

	1990	76	50	12	138	10	12	2	24
	1995	57	58	22	137	10	13	6	29
	2000	43	54	30	127	8	10	8	26
	2004	40	40	30	110	10	8	12	30
Venezuela	1985	22	29	2	53	0	2	0	2
	1990	19	39	4	62	3	3	1	7
	1995	7	25	9	41	0	7	6	13
	2001	4	33	10	47	1	6	7	14
TOTAL for four countries	1985	108	87	23	217	8	5	9	21
	1990	130	126	37	292	17	19	18	54
	1995	88	116	63	267	18	24	25	67
	2001	67	124	73	264	17	21	26	64

There was also an increase in the total number of scientists involved in biotechnology research in Cameroon, from 3 in 1985 to 18 in 2001. The number of Ph.D. scientists reached a maximum of 9 scientists in 1990 and decreased to 6 in 2001. The number of M.S. scientists remained fairly stable at 4-5, while the number of scientists with a B.S. degree increased from 4 in 1995 to 7 in 2001. While these numbers represent an overall increase in the pool of knowledgeable people, the financial resources attributed to plant breeding did not increase at the same pace (Table 5).

Table 5: Financial resources and resources per researcher (1993 international PPP dollars)

Country	Year	Financial resources Total	 Plant breeding	 Biotechnology	Financial resources per researcher Plant breeding	 Biotechnology
		1993 international PPP dollars, thousands			**1993 international PPP dollars**	
Cameroon	1985	20,212	4,589	231	169,967	76,942
	1990	5,880	905	403	18,091	23,687
	1995	15,977	1,030	208	26,399	10,383
	2001	27,145	1,050	218	27,643	12,122
Kenya	1985	15,138	9,096	736	232,640	171,242
	1990	32,070	18,068	2,837	426,123	197,673
	1995	14,210	7,927	1,337	159,825	252,263
	2001	13,629	6,773	1,634	129,744	259,380
Philippines	1985	12,059	8,982	280	91,653	23,317
	1990	18,281	10,907	1,371	79,038	57,135
	1995	26,209	10,952	2,562	79,940	88,333
	2000	57,068	14,896	6,608	117,293	254,172
	2004	149,219	21,619	7,808	196,532	260,275
Venezuela	1985	3,028	615	7	11,602	3,628
	1990	3,753	546	16	8,804	2,220
	1995	20,871	1,688	117	41,182	9,033
	2001	8,749	1,384	79	29,445	5,618

In fact, there was a significant drop in plant breeding expenditures from a little over 4 million 1993 international dollars in 1985 to 905,000 in 1990, with the budget remaining around this level in the following years, showing only a slight increase over time. On the other hand, biotechnology-targeted resources peaked briefly in 1990 (at 403,000 1993 international dollars) and otherwise remained fairly stable around 218,000 1993 international dollars.

The sharp fall in resources for plant breeding seen around 1990 (Table 5) reflects the economic crisis experienced by Cameroon. The GDP fell almost 4% annually during the period 1987-1994, with a record 7.8% decrease rate seen in 1988. These circumstances forced the government to reduce its overall expenditures, thereby affecting all programmes, including the crop breeding budgets. These decreases in plant breeding resources, coupled with the observed increases in the number of scientists, meant that the expenditures per researcher decreased drastically after 1985. Indeed, expenditures per researcher in both plant breeding and biotechnology decreased significantly from 1985 to 2001. Expenditures for plant breeding fell from 169,000 to 27,000 1993 international dollars per researcher, while those for biotechnology fell from roughly 76,000 to 12,000. Although there is no available standard that use to judge this level of per-researcher expenditure, policy makers in Cameroon should probably be concerned about the sustainability of research and development (R&D) activities with such a limited per researcher budget.

Effort Allocation by Crop

In 2001, the largest shares across all organisations surveyed in Cameroon were allocated to germplasm enhancement (35%) and line evaluation8 (39%), while line development accounted for 26% of the total effort dedicated to plant breeding. In real terms, public sector organisations in Cameroon invested 0.37 million in germplasm enhancement, 0.27 million in line development and 0.41 million for line evaluation. The total investment across all studied activities was 1.05 million 1993 international dollars. The distribution of expenditures by crop (Figure 1) shows that in 1985 the four crops and crop groups with the highest emphasis in terms of expenditures were maize, rice, vegetables and fruits, and roots and tubers. By 2001, however, the highest expenditures were on vegetables and fruits, followed by fiber crops (cotton), roots and tubers, and maize.

Sources of Germplasm

the highest share of germplasm source (55%) was represented by local germplasm banks, which included local and national sources. The public sector tapped unto bilateral agreements, CGIAR and germplasm evaluation networks, accounting for 38% of germplasm sources. The private sector supplied only 5% of all sources of germplasm to the public sector in Cameroon.

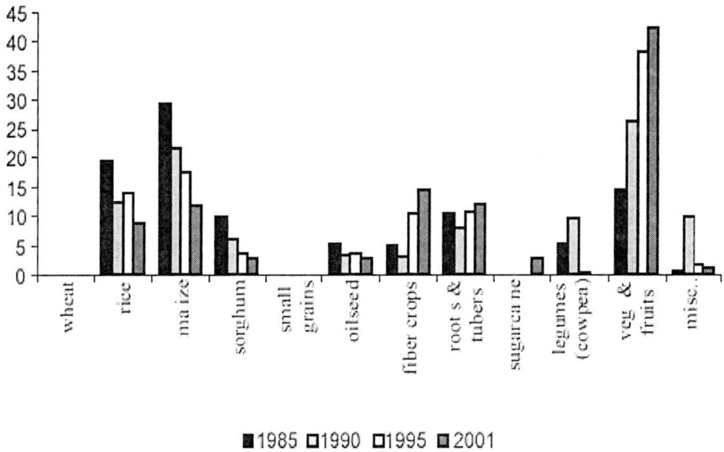

Source: Plant breeding and biotechnology capacity survey

Figure 1: Cameroon—Plant breeding budget by crop (as a percentage of total breeding budget)

Kenya

Output in Terms of Varieties or Lines Released

The number of varieties released by the public sector in Cameroon decreased over time, from 37 varieties released in 1985 to 6 varieties in 2001. In 1985, the public sector in Cameroon released varieties of all main crops except for sorghum, millet, vegetables and fruits. In contrast, in 2001, the public sector in Cameroon only released varieties of rice, fiber crops, vegetables and fruits. Although the number of varieties released can be used to describe the effectiveness of outputs produced by the public and private sectors in a given country, the percentage of adoption (or area planted) of varieties released by both sectors would be a much better indicator, as it would speak to the quality of the released varieties. However, it is difficult to consistently obtain such information in many countries worldwide.

Institutional Information

At least ten institutions in Kenya, including both public and private, have ongoing programmes for plant breeding research. These organisations include one public research institution, four public universities, and five private seed companies (Table 3). The oldest and largest among them is the Kenyan Agricultural Research Institute (KARI), which has been involved in plant breeding for more than 85 years. In 1999, the Agricultural Science and Technology Indicators (ASTI) database estimated that KARI accounted for more than half of the national agricultural research expenditure and staff. A major player representing the private seed sector in Kenya is a producers' organisation called the Seed Traders Association of Kenya (STAK), which was established in 1982. STAK, through the auspices of the Kenya Plant Health Inspectorate Services (KEPHIS), is licensed to produce, process and/or distribute seeds in Kenya. STAK members are currently involved in the research and multiplication activities of most important crops in this country, such as maize, wheat, beans, sunflower, vegetables and fodder. The member seed companies, which currently comprise 20 private firms, contribute nearly 90% of Kenya's total formal seed sector. Out of the 20 firms, only four support R&D activities (namely breeding and varietal release). Most biotech research in Kenya especially that focused on maize is carried out by the multinational firms.

Kenya is one of the few African countries with an ongoing research agenda in biotechnology. The IFPRI Next Harvest 2002 study documented six GM projects led by KARI, aimed at improving insect, herbicide and virus resistance in maize, cotton, sweet potatoes, and potatoes. However, the overall budget and resources for biotechnology have declined since 1990 while the number of researchers has increased, signaling detrimental decreases in per-researcher resources.

Resources Dedicated to Plant Breeding and Biotechnology

Table 4 shows the total number of research staff members employed in plant breeding and biotech (FTE employees) in Kenya from 1985 to 2001. Overall, the total number of research staff members increased in both areas, from 39 to 52 and from 0.3 to 6.3 for plant breeding and biotechnology, respectively. In contrast, the total financial resources for public and private plant breeding in Kenya declined from 1985 to 2001, at almost the same rate as the decline in this country's overall agricultural research budget (Table 5). Funding for plant breeding increased to 18 million 1993 international dollars in 1990, but decreased significantly thereafter, until the funding in 2001 was below that seen in 1985. In contrast, the funding for biotechnology R&D steadily increased from 0.73 million in 1985 to 1.6 million in 2001.

The 25 percent decline in the level of financial resources allocated for plant breeding from 1985 to 2001, concurrent with increases in the number of research staff members (particularly at the Ph.D. level) reduced the level of per-researcher expenditure. Expenditures per researcher decreased almost in half from 1985 to 2001, whereas biotechnology saw a small increase in funding. The serious decline in per-researcher expenditure for plant breeding is likely to limit Kenya's research capabilities in this area. Total allocations by institutions participating in agricultural research for plant breeding and biotechnology R&D increased from 1985 to 1990, and decreased thereafter until the 2001 levels were similar to those of 1985. A similar trend was seen in the budget for plant breeding. These fluctuations are consistent with ASTI findings for overall trends in agricultural R&D expenditures during this period.

Effort Allocation by Crop

In the case of fund allocation in Kenya, line evaluation comprised the largest share (55%) of investments, whereas germplasm enhancement and line development represented 23% and 22% of those expenditures in plant breeding, respectively. These trends show that research efforts in Kenya during the study period were geared more toward adapting existing lines rather than developing new seeds. As seen in Figure 2, the percent shares of resources dedicated to maize, oilseeds, forages, and wheat decreased in Kenya over the study period. The percent shares of roots and tubers, fiber crops, rice, and other small grains remained fairly constant, but represented a relatively small part of the overall breeding budget. Finally, the percent shares of sorghum/millet, other legumes and vegetables and fruits increased over time.

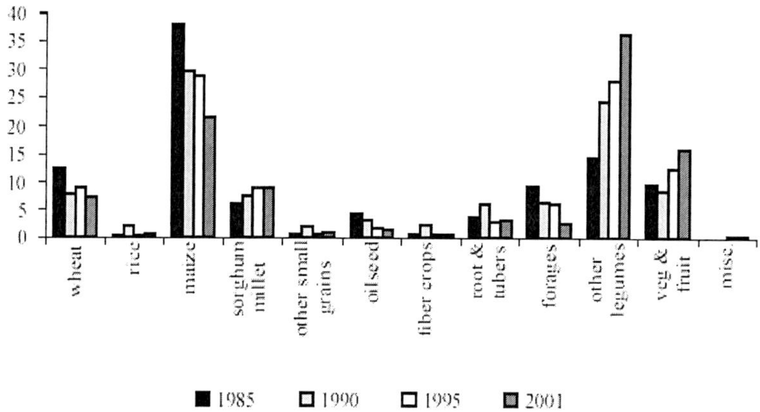

Source: Plant breeding and biotechnology capacity survey

Figure 2. Kenya—Plant breeding budget by crop.

Contrary to the case in Cameroon, Kenya had a functional private sector investing in plant and/or biotechnology R&D, allowing us to separate effort allocation by type of institution. Figure 3 shows germplasm enhancement, line development and line evaluation differentiated into their public and private sector effort allocations. As expected, the Kenyan private sector invested the most in line evaluation and the least in germplasm enhancement during the study period. Unexpectedly, the public sector targeted an even higher share of its investments into line evaluation activities, compared to the private sector. We speculate that this is probably due to the ongoing evolution of the plant breeding system in Kenya, where the public sector has traditionally dominated and the private sector is just starting to enter the market.

Sources of germplasm

The sources of germplasm for plant breeding efforts in Kenya came mainly from local and national germplasm banks (58%), with a smaller percentage coming from CGIAR gene banks (24%), public organisations in industrialised countries (9%), and the private sector (8%). The survey results indicate that private multinational companies in Kenya drew 100% of their sources from local germplasm banks during the study period, whereas private local companies were more diversified, with 51% of their germplasm coming from local and national germplasm banks and another 38% from public organisations in industrialised countries.

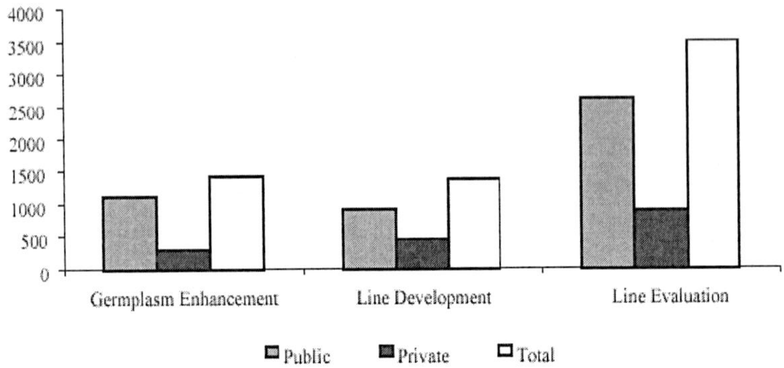

Source: Plant breeding and biotechnology capacity survey

Figure 3. Kenya—Distribution of plant breeding emphasis by type of institution.

The Philippines

Institutional Information

The Philippines have a significant number of institutions dedicated to agricultural research. The survey contained consistent data across five public and private sector institutions, representing 18-80 and 5-26 years of experience in plant breeding and biotechnology, respectively (Table 3).

Human and Financial Resources Dedicated to Plant Breeding and Biotechnology

Table 4 shows the distribution of human resources for the plant breeding and biotechnology organisations surveyed in the Philippines. The total number of scientists involved in plant breeding increased from 1985 to 1990, when it peaked at 138 researchers, and decreased thereafter to 110 in 2004. In contrast, the total number of scientists involved with biotechnology more than doubled over the study period, from 12 in 1985 to 30 in 2004. Examination of the distribution by each scientist's degree for plant breeding reveals that the number of B.S. scientists decreased, the number of M.S. scientists increased slightly, and the number of Ph.D. scientists increased significantly, from 9 in 1985 to 30 in 2004. The number of scientists in biotechnology increased across all scientists' degrees, with the highest increase seen at the Ph.D. level where it increased from 1 scientist in 1985 to 12 in 2004. This pattern may suggest that the Philippines pursued advanced techniques through investments in human resources during the study period. There is some concern regarding the lack of a new generation of scientists within the R&D system to take over for retiring scientists. The conducted a follow-up survey of the age distribution of scientists in the Philippines at the surveyed institutes10. The results of this exercise (Figures 5a-c) show that the age distribution of plant breeding and biotechnology researchers is indeed increasing over time. Although the number of Ph. D. scientists in the Philippines has increased significantly over the period, the hiring of young scientists has decreased and altogether stopped by 2004. As most scientist are now over age 45, it is important that policy makers in the Philippines develop and implement plans towards training a new cadre of scientists.

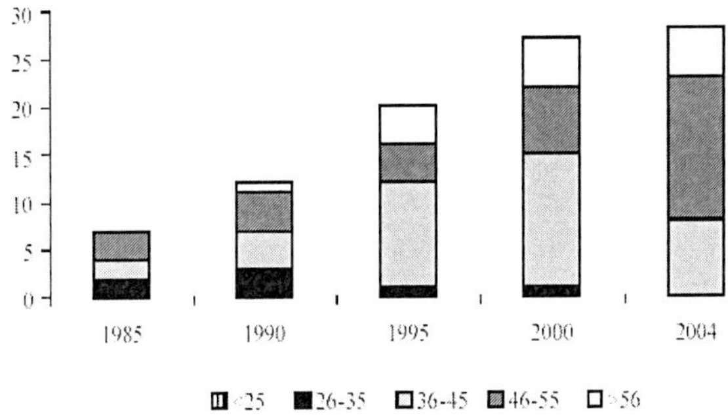

Source: Plant breeding and biotechnology capacity survey

Figure 5a. The Philippines—Number of Ph.D. researchers by age distribution.

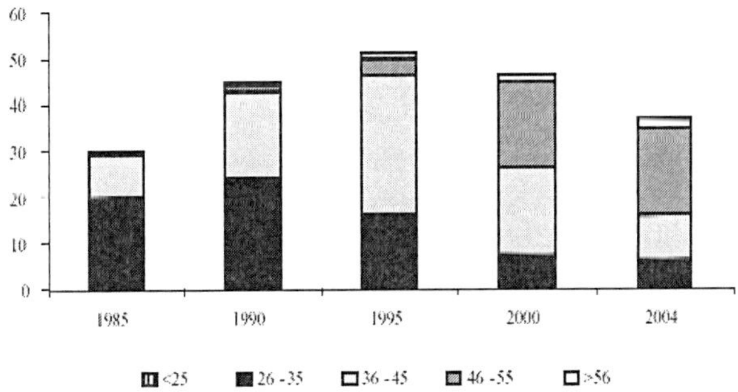

Source: Plant breeding and biotechnology capacity survey

Figure 5b. The Philippines—Number of M.S. researchers by age distribution.

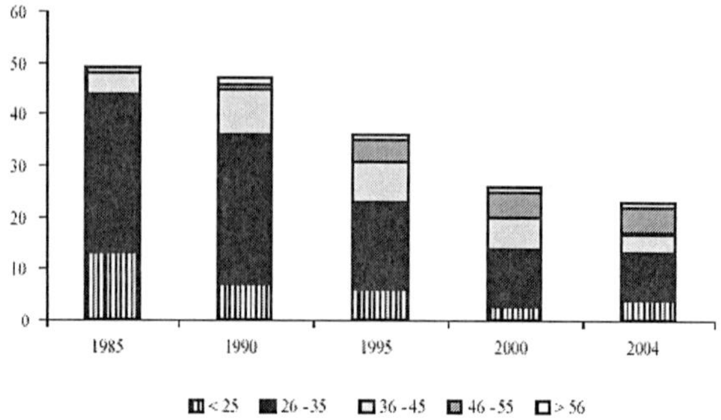

Source: Plant breeding and biotechnology capacity survey

Figure 5c. The Philippines—Number of B.S. researchers by age distribution

Table 5 shows data on the financial resources available for plant breeding and biotechnology efforts in the Philippines from 1985 to 2004. The total financial resources for the surveyed institutes increased for both plant breeding and biotechnology. The total budget for plant breeding increased from 8.9 million international dollars in 1985 (given in 1993 dollars) to 21 million in 2004, while that for biotechnology increased from 0.28 million international dollars in 1985 to 7.8 million in 2004. These increases in human and financial resources for both plant breeding and biotechnology indicate that the government of the Philippines and other investors viewed investments in R&D as promising over this period. However, development of technologies to address productivity constraints in the agricultural sector is still needed. Table 5 also shows that the financial resources available per researcher increased in both plant breeding and biotechnology from 1985 to 2004. The rate of increase was higher for biotechnology during this period, with per-researcher funds in plant breeding nearly doubling from a 1985 level of 91,000 1993 international dollars to 196,000 in 2004, while those for biotechnology increasing nearly 10-fold from 23,000 international dollars in 1985 to 260,000 in 2004.

A given country's investments in financial and human resources for research activities should be evaluated in terms of their relative contribution to improving societal welfare and other metrics. One method of evaluation is to contrast the investments made in a particular crop with their relative contribution to the economy in terms of value of production. This metric is limited in that it tends to mask social impact of the crop and discriminate against intensive production crops such as vegetables and/or orphan crops, while not giving an estimation of potential future value. Therefore, this metric is combined with others to properly assess priority setting and post-activity evaluation processes, in both private and social terms. This metric in order to showcase the utility of such analyses in countries that have a fairly mature market economy, such as the Philippines. To start assessing the congruency of plant breeding spending by crop with the actual importance of those crops in the economy, compare the percentages spent for the main crops/groups with their participation in terms of value of crop production. Ideally, this comparison would be done with the same base year. Since this information is not available, examine the value of production calculated with FAO 1986-91 average international prices in dollars, versus the plant breeding expenditure in 1993 international dollars. Although these figures have different base years and are thus not strictly comparable, get a rough estimate by comparing the percentages rather than absolute values (see Figures 6a and 6b).

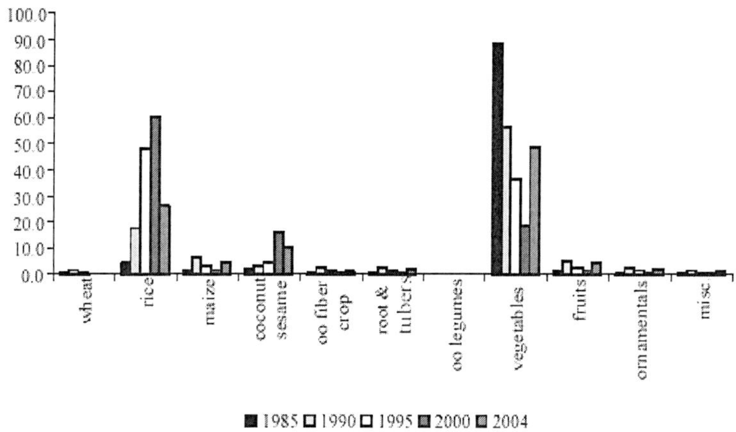

Source: Plant breeding and biotechnology capacity survey

Figure 6a. The Philippines—Plant breeding budget by crop

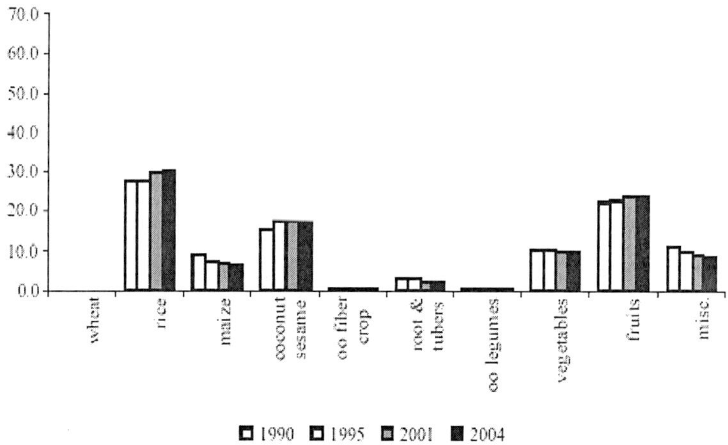

Source: Plant breeding and biotechnology capacity survey

Figure 6b. The Philippines—Value of crop production

Our analysis reveals that the most important crops in the Philippines in terms of value of production during the examined time period were rice, coconut, bananas, fresh vegetables, and tropical fruits. These five groups accounted for more than 70% of the value of crop production in 2004. Figures 6a and 6b show that there is not a smooth congruency between the economic importance of a given crop or group of crops and the corresponding expenditure in plant breeding.

For example, the production value share of rice in the Philippines (crops only) was around 30% over the past 20 years, whereas the FAO survey results indicate that plant breeding expenditures in rice oscillated between 18% and 60% of total expenditures in plant breeding during this period. Plant breeding expenditures in vegetables showed similar percentages, but their participation in the crop economy was only around 10%.

Effort Alocation by Crop

Allocation of plant breeding efforts in the Philippines over the study period was relatively balanced across germplasm enhancement (28%), line development (36%) and line evaluation (36%) activities, as shown by the financial resources invested in these areas. In terms of actual resources invested, the surveyed organisations invested 5.9 million dollars in germplasm enhancement, 7.7 million in line development, and 7.6 million in line evaluation, for a total plant breeding investment of 21.3 million dollars (all expressed as 1993 international dollars).

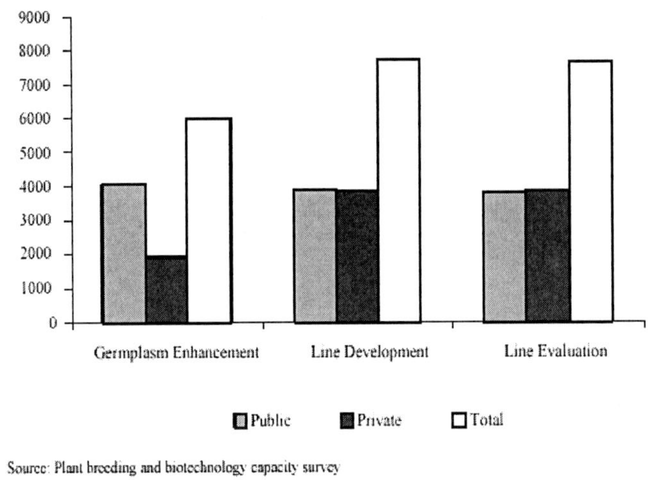

Figure 7. The Philippines—Distribution of plant breeding emphasis by type of institution, 2004.

Figure 7 shows these data disaggregated by plant breeding emphasis and by type of institution in the Philippines. As expected, the private sector tended to focus more on line development and evaluation, and less on germplasm enhancement, compared to the public sector (however, private investments were made in all three areas). This is congruent with the suggestion made by Falck-Zepeda and Traxler, who proposed that private sector plant breeding institutions tend to concentrate their efforts more on applied rather than basic research.

Venezuela

Institutional Information

Although five institutes in Venezuela were identified and surveyed, this report includes only those three for which complete financial data were available. Even though the social importance of agriculture is small within the Venezuelan economy, due to the abundant resources provided by oil production, see from Table 3 that the organisations in the survey have made longterm commitments to plant breeding and biotechnology efforts.

Human and Financial Resources Dedicated to Plant Breeding and Biotechnology

As shown in Table 4, the total number of scientists involved in plant breeding in the surveyed Venezuelan institutes decreased slightly from 53 in 1985 to 47 in 2001. In contrast, the total number of scientists involved with biotechnology increased seven-fold, from two in 1985 to 14 in 2001. Examination of the distribution by each scientist's degree reveals that in the case of plant breeding, the number of B.S. scientists decreased over the study period, the number of M.S. scientists increased slightly, and the number of Ph.D. scientists increased significantly from two in 1985 to 10 in 2001. In biotechnology, the number of both M.S. and B. S. scientists in biotechnology decreased, while there was a significant increase at the Ph.D. level. This pattern seems to suggest that Venezuela pursued technical advancement in plant breeding and biotechnology through investments in human resources during the survey period. The total financial resources for both plant breeding and biotechnology among the surveyed institutes, expressed as 1993 international dollars, increased overall between 1985 and 2001, although a small decline was seen in 2001 (Table 5). The investments were relatively small in comparison to those of the other surveyed countries.

The total budget for plant breeding increased from 615,000 1993 international dollars in 1985 to 1.38 million in 2001. The change in biotechnology spending was much larger in relative terms, as it increased from 7,000 dollars in 1985 to 79,000 dollars in 2001. This small level of investment reflects the relatively low value of agriculture in the Venezuelan economy. However, the increases in both human and financial resources for both plant breeding and biotechnology among the public and private sector organisations in Venezuela suggest that these areas are gaining increased attention and investors are seeking to address productivity constraints in the agricultural sector. Table 5 also shows that the financial resources available per researcher increased in both plant breeding and biotechnology from 1985 to 2001, with the rate of increase in plant breeding exceeding that in biotechnology. The average financial resources per plant breeder totaled 11,602 (in 1993 international dollars) in 1985, and increased by 153% to $29,445 in 2004. In contrast, the per-researcher funding for a biotechnologist increased by only 55%, from $3,628 in 1985 to $5,618 in 2004. Although there are no real standards for comparing these indicators, the level of financial resources per researcher in Venezuela was the lowest during this period compared to any of the three other countries studied herein.

Effort Allocation by Crop

The emphasis in Venezuelan plant breeding was distributed fairly evenly between line evaluation (40%) and line development (37%), with germplasm enhancement ranked somewhat lower (23%). As shown in Figure 8, where further disaggregate these data by type of institution, the private sector invested a higher share of its resources into line evaluation and placed the least emphasis on germplasm enhancement. In contrast, the public sector invested the highest share of its resources in line development followed by line evaluation, with the lowest level of investment targeted toward germplasm enhancement.

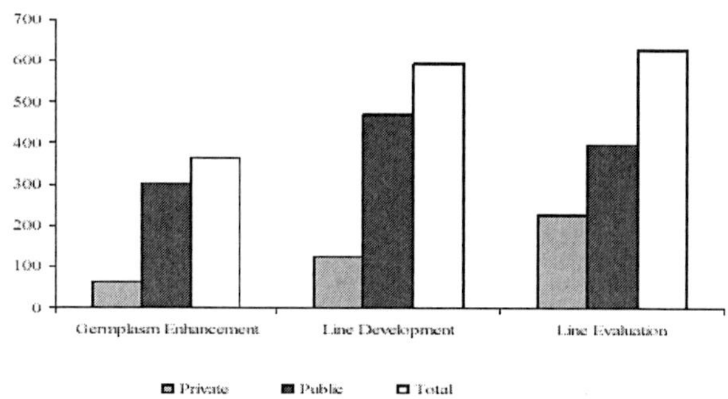

Source: Plant breeding and biotechnology capacity survey

Figure 8. Venezuela—Distribution of plant breeding emphasis by type of institution, 2001.

Output in Terms of Varieties

Most of the varieties released in Venezuela during the study period were produced by the public sector, although varietal release by the private sector increased somewhat after 2001. The bulk of varieties released by the private sector were developed from germplasm produced elsewhere. The production of final varieties increased slightly from 56 in 1985 to 60 in 2001. In total, over 175 public and private varieties were released between 1985 and 2001. The majority (52%) of released varieties were cereals (e.g. wheat, rice, maize, sorghum and millet), followed by roots/tubers and legumes (21%) and oil and fiber crops (6%). The balance was distributed among vegetables, fruits, and other miscellaneous crops.

All plant breeding and biotechnology R&D programmes must assess whether they are addressing the needs of their stakeholders. One potential indicator of the appropriateness of a given R&D programme is the congruence between the share (%) of total value of production per crop and the share (%) of investments in a particular crop or crop group. This relationship between R&D investments and crop priorities can provide valuable insight into investment decisions. In Figure 9, present an estimation of the congruency between the value of production and investments in plant breeding R&D in Venezuela. This indicator of congruency or parity is a common method for assessing the allocation of research resources. A value of 1 (denoted by the horizontal thick line) represents a perfect congruency between shares. Values above 1 indicate that the investments in a particular crop for a given year are much higher than the crop's share of value of production in that year. This situation represents an over-investment in a particular crop or crop group. Values less than 1, on the other hand, represent under-investment. We calculated these ratios only for Venezuela because the value of production data was incomplete for the other three countries.

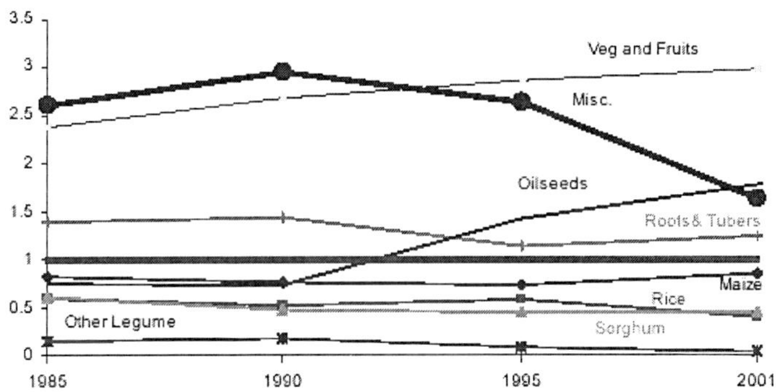

Source: Plant breeding and biotechnology capacity survey

Figure 9. Venezuela—Congruencies between value of production and plant breeding emphasis

We can see from Figure 9 that during the study period, Venezuela over-invested in vegetables/fruits, miscellaneous, and roots and tubers, whereas they under-invested in cereals and other legumes. In the specific case of oilseeds, Venezuela under invested in plant breeding R&D for these crops in 1985, then moved towards over investing by 1990. We would not expect to have perfect congruency in any country, especially as significant stakeholder, institutional and governance issues may be involved in the decision making process. In that sense, an ideal scenario would be one in which R&D investments converge toward the value of production, at least for this indicator. As indicated previously, congruency analysis should be part of a broad priority setting and post-activity evaluation process that includes other metrics to properly assess investments in terms of a country's portfolio of crops, and their traits and productivity constraints.

POLICY IMPLICATIONS AND RECOMMENDATIONS

General Policy Interventions

Tables 6a-e show the modified version of Trigo's table that used to incorporate some of the innovative influencing policies discussed by Furman, Porter and Stern into the Fuglie and Pray frameworks. We modified both approaches to explain PGRFA, plant breeding and biotechnology as distinctive innovative activities.

Table 6a: Determinants of innovation, R&D, and technology transfer

Innovation- R&D and technology transfer determinants	Policies
General state of the economy	Macroeconomic stability
	Public infrastructure
	General education and training
	Development of capital and insurance markets
Size of input markets	Market share of state-owned enterprises
	Trade restrictions on inputs
	Price interventions in input or product markets
Technological opportunity and the	Public investments in agricultural research and education
cost of research inputs	Registration and testing requirements for new seed and agricultural chemicals
	Biosafety requirements for biotechnology field trials
	Public subsidies for private research
Appropriability	Intellectual property laws
	Technology licensing requirements and regulations affecting technology imports
	Competitiveness and antitrust policies

The value of discriminating countries into distinct polity situation categories based on specific indicators and policies influencing innovation (as done in Trigo's. approach) is that it makes it possible to qualitatively map countries against the general policy interventions. Ideally, if enough data were available, it would be possible to quantitatively determine the relative importance and value of each factor or determinant in explaining innovation, pinpointing gaps and limitations within the innovation process. In fact, this is the process followed by Furman *et al.* in their model for the Organisation for Economic Co-operation and Development (OECD) member countries.

An important lesson from the typology and classification procedures described by Trigo is that countries that fall into the category "setting the stage for using biotechnology products" may need to concentrate more on developing the basic infrastructure for performing agricultural research while using 35 and conserving PGRFA and for conventional plant breeding, rather than trying to develop new biotechnology capacities. One of the potential outcomes of these countries' investments in their "common innovative infrastructure" is the creation of the basic platforms needed to take advantage of spillovers from innovator countries or from those in the "improving the efficiency" category. Of the four countries in the survey, Cameroon fits into this category.

Table 6b: Determinants of innovative capacity for developing economies in the FAO survey—common innovation infrastructure and contributing outcome factors

Determinants of innovation	Indicators	Unit	Philippines	Kenya	Venezuela	Cameroon
Overall scientific capacity	Researchers in R&D per million people	Number per million	156	?	189	?
	Technicians in R&D per million people	Number per	million	22	?	32 ?
	Total Gross Expenditure on R&D as percent of GDP	%	0.11	?	0.39	?
	Public expenditure on education as % of GDP-2003	%	3.2	6.9	?	3.8
	Average publications scientific/tech. journals 1993-20011	Number	161	274	457	72
Intellectual property	All patents granted in the US to countries 1977-2005	Number	276	57	604	0
	Utility patents granted in the US to countries 1963-2005	Number	284	58	658	2
	Patents granted to		na			
	residents	Number		0	14	11
	non-residents	Number		33	742	1,160
	(Source WIPO)	(Year)		(2001)	(2000)	(2003)
	Plant variety protection legislation	Year in force	1992	1977, amended in 1991.	1998	2001
Economy-wide issues	Trade and Development Index 2005	Index number (Global rank)	478 (95)	359(84)	440(70)	248(100)
	Average population 1991-2003	Millions	71	28	23	14
	Real Gross Domestic	2000	3,781	1,072	6,019	1,965

Product per capita	international $				
Average Gross Domestic Product 1984-2004	Millions, 2000 international PPP dollars	245,618	26,921	125,393	24,893
Average labor force 1984-2004	Thousands	25,611	10,865	8,092	4,757
Foreign direct investment, net inflows	% of GDP	1.76	0.25	2.78	–0.06

Table 6c: Determinants of innovative capacity for developing economies in the FAO survey—common innovation infrastructure and contributing outcome factors (continued)

Determinants of innovation	Indicators	Unit	Philippines	Kenya	Venezuela	Cameroon
Overall technological and industry capacity	Average high technology exports 1991-2004	% of manufactured exports	56	4	3	2
	Average information and communication technology investments 2000-2004	% of GDP	5.5	3.0	4.3	4.9
	Industry value added	% of GDP	33	18	50	25
Governance	Voice and accountability	Index 2006	0.02	–0.3	–0.47	–1.17
	Political Stability	Index 2006	–1.23	–1.07	–1.13	–0.59
	Government Effectiveness	Index 2006	–0.17	–0.72	–0.98	–0.70
	Regulatory Quality	Index 2006	–0.20	–0.25	–1.11	–0.77
	Rule of Law	Index 2006	–0.67	–1.01	–1.22	–1.12
	Control of corruption	Index 2006	–0.58	–0.58	–1.0	–0.98
International Treaties	WTO member	Y= Yes	Y	Y	Y	Y
	Cartagena Protocol on Biosafety	S=Signed R=Ratified	S/R	S/R	S/R	S/R

In contrast, countries in the category of "improving the efficiency and products of agricultural research through the use of biotechnological tools" have significant capacity to perform agricultural research but may not yet be advanced enough to create biotechnology innovations by themselves. These countries may be readily able to take advantage of direct spillovers from other countries, and to do adaptive research supporting the use of technology developed elsewhere. The capacity of these countries may be influenced by their market size. The range of factors and tools available to countries in this category increase significantly as more research and development opportunities arise; this, in turn, changes the sophistication level of their innovative system. Three of the four countries in the sample fall into this category, namely the Philippines, Kenya and Venezuela.

Table 6d: Determinants of innovative capacity for developing economies in the FAO survey—cluster-specific

Determinants of innovation	Innovative indicators	Unit	Philippines	Kenya	Venezuela	Cameroon
Conservation and enhancement of PGRFA	Average investments in germplasm enhancement 1985-2001	Thousands 1993 international dollars	5,976	1,426	363	224
	Average investments in germplasm enhancement relative to total budget institutions, four observations 1985-2001	%	4.0	1.9	4.2	0.82
Conventional plant breeding	Capacity level	Subjective	xx	xx	x	x
	Average full time equivalent number of scientists in plant breeding 1985-2001	Number of FTE	122	46	51	39
	Average investments in plant breeding 1985-2001	Thousands 1993 international dollars	13,471	10,446	1,058	1,894
	Number of PVP registrations	Number				
Biotech capacities	Capacity level	Subjective	xx	x	x	x
	Biotech patents	Number	0	7	?	?
	Number of scientists in biotech R&D	FTE, average 1985-2001	24	5	9	15
	Average investments in biotech R&D	1993 international PPP dollars, thousands	3,726	1,636	55	265
	Field trials—Total	Number	7	4	0	0
	Field trials—university/ NARs/NAROs	Number	2	1	0	0
	Events approved (number of crops) for food and/or feed consumption	Number	37 (7)	0(0)	0(0)	0(0)
Overall crop improvement cluster relative strength	Percentage of total plant breeding R&D performed by public sector	%	57	60	92	100
	Percentage of total plant biotechnology R&D performed by public sector	%	57	60	92	100

Finally, those countries in the category of "building capacities to develop biotechnology-based innovations" have achieved a distinct capacity to internally generate innovations from basic concepts at the discovery phase of research, and develop them all the way to commercial applications. It is worth noting that none of the four countries in the survey fall into this category. In fact, Trigo identified only nine developing countries in this category: China, Brazil, India, Mexico, Argentina, South Africa, Thailand, Russian Federation and Indonesia.

Table 6e: Determinants of innovative capacity for developing economies in the FAO survey—links, networks and technology transfer capacity

Determinants of innovation	Innovative indicators	Unit	Philippines	Kenya	Venezuela	Cameroon
Market size	Arable land	Category	Med.	Med.	Med.	Med.
	Value added Agriculture	Percentage of GDP	16	26	5	39
	Internal seed market	Million US$	–	50	–	–
	Number seed firms	Number	At least 19	18	–	–
	Seed imports	Million US$ (FOB)	8	11	13	?
Strength of the private sector	Domestic credit to private sector 2004	% of GDP	34.8	26.8	11.2	9.9
	Ease of doing business	Global rank 2005	121	80	144	147
	Cost of business startup procedures (average 2001-2003)	% of GNI per capita	21	51	15	182

Current policy interventions in Cameroon are focused towards improving the common national innovative infrastructure, facilitating basic research into the use and conservation of PGRFA, and promoting conventional plant breeding. In this sense, the current state of development of the innovation system may indicate higher payoffs to investments in more conventional plant biotechnology, and perhaps the adaptation of advanced biotechnologies that have been developed elsewhere. As this is a dynamic process, these steps will enable the country to improve its innovation system and allow it to move forward 36 to the next phase (that of "improving the efficiency..."), where greater emphasis will be put on the use and development of new biotechnology innovations. For Kenya, Venezuela and the Philippines, which are classified under "improving the efficiency...," specific tools need to be identified to effectively address the cluster-specific gaps and limitations affecting crop improvement. Greater emphasis should be placed on the abilities of these countries to evaluate, adapt, and use advanced biotechnology innovations. Nevertheless, the capacity to use and conserve PGRFA remains as critical as it is in the other categories, because biotechnology innovations cannot advance in the absence of previous ("intermediate") PGRFA innovations.

The Philippines are at a more advanced stage of development than the other two studied countries in that group and are moving forward in terms of producing GM biotechnologies within their innovation system. The Philippines have invested significant resources in both the common national innovation capacity and in cluster-specific groups for crop improvement. However, additional resource investments may be needed to improve the links between national clusters and other international groups, as well as technology transfer mechanisms. The data evaluated in this study seem to indicate that this country has a significant common national innovation structure, and has made important investments in plant breeding clusters. If Venezuela decides to make future investments in the biotechnology cluster, policy makers should probably address the question of aligning their investments with crop priorities. Countries may be classified according to their capacity to perform biotechnology innovations. An example of this is the approach developed by Trigo, wherein countries are categorised according to their innovative capacity and ability to tap into capacities developed elsewhere. The four categories described by Trigo are: (1) importers of technology—non-selective; (2) importers of technology–selective; (3) tool users; and (4) innovators. Among the four countries examined herein, Trigo (Table 5) classified Venezuela and Kenya as "selective importers of technology with a medium market" and the Philippines within the category of "tool users with a medium market".

Potential Country-Specific Policy Interventions

Based on the analysis of the available data for the four studied countries (See Table 7), propose a set of country-specific policy interventions derived from the survey data, IFPRI's analysis and experience, and particularly the authors' in-country experience with plant breeding and biotechnology issues.

Table 7: Comparative indicators among countries

Indicator	Cameroon	Kenya	Philippines	Venezuela
Population 2003, million hab.	15	31	78	25
Ag GDP, as a percentage of total GDP	44	19	15	4
Number of Ph.D. plant breeders in most recent year, FTE	13	20	30	10
Number of Ph.D. plant breeders in 1985, FTE	8	4	9	2
Key crop groups for breeding, number	3	5	1	5
Number of surveyed institutes	5	10	5	3
Budget for plant breeding, annual average	1,894	10,446	13,471	1,580
Budget for plant breeding, most recent year	1,050	6,773	21,619	1,384
Budget per researcher, most recent year	27,643	129,744	196,532	29,445
Budget for biotechnology, most recent year	265	1,636	3,726	55
Number of varieties released, most recent year	6	45	78	60
Number of varieties released per year, average	25	31	43	48
Value added as a percentage of GDP	39	26	16	5

Cameroon

Two of the most pressing policy issues in Cameroon are the low overall funding for plant breeding and biotechnology, and the low available funds per scientist. As shown inthe analysis, Cameroon's budget for plant breeding and biotechnology has declined greatly since 1985, and it has the lowest rate of funding available per researcher among the studied countries. This would seem rather counterintuitive, given the economic importance of the agricultural sector and the growing value of agriculture in this country, as measured by its share of total GDP. An additional concern in Cameroon is the relatively low crop coverage by scientists in terms of crop investment and priority preferences and outputs (varieties released). The 13 FTE scientists available for plant breeding in Cameroon are spread over five institutes and work on multiple crops. Notably, in the earlier years studied, Cameroon pursued a more diversified breeding strategy that included wheat, maize, rice, and fiber crops, three of which (maize, rice, and wheat) were basic global food security crops. By the end of the studied timeframe, however, only rice, fiber crops, vegetable and fruits varieties were being released. In addition, crop selected for research in development and posterior release in Cameron (as in Table 7a), should be contrasted with Cameroon's staple foods, which include corn, millet, peanuts, yams, cassava, and plantains, as well as their major export commodities, such as cocoa beans, coffee and cotton.

Declining resources and a shift in crop prioritisation is likely to curtail both breeding and biotechnology programmes in Cameroon. One way to expand the knowledge base and research in this country might be to implement incentives aimed at encouraging the private sector to become involved in meeting the needs of farmers. This may be possible for crops having significant investments at the international level (e.g. maize), provided there is a market and cash base to support further market development in Cameroon. However, as shown by the analysis of the national innovating capacity, there are only limited opportunities for national private sector investment in plant breeding and biotechnology in Cameroon. This may represent an opportunity for the government of Cameroon to support governance and targeted support of private sector investments in the country, which will indirectly aid the development of agricultural technologies in general.

Kenya

A major consequence of Kenya's expected future population growth is the need for crop productivity improvements, including development of new varieties or hybrids through conventional plant breeding and other less traditional approaches. The experience worldwide is that in cases of stagnant productivity, farmers tend to expand production to marginal lands in order to meet expanding demand and food security needs and this may well be the case for Kenya. However, these developments needs may conflict with other land use purposes (conservation, game reserves, etc.), making it important to also examine the possibility of significantly increasing the productivity of the farmland that is already in use. The analysis shows that the Kenyan budgets for plant breeding and biotechnology decreased from 1990 to 2001. The largest (although decreasing) investments were made for maize and legumes; in 2001, legumes represented the largest share of the national plant breeding budget, followed by maize, with a major emphasis on line evaluation over germplasm enhancement and line development. Over the study period, the plant breeding capacity, as measured by the number of Ph.D. increased from 4 to 20, with these scientists spread over 10 institutes. Among the four studied countries, Kenya was second only to the Philippines in total number of Ph.D. scientists in plant breeding. Kenya also invested a significant amount of financial resources in plant breeding over the period of the survey; this built a strong foundation for variety development, resulting in the release of 45 releases by the public sector in 2001, which is well over the period-average of 31.

One potential problem for the plant breeding system in Kenya is that the available funds per researcher are approximately $130,000 (1993 international PPP dollars) per year in 2001. In the authors' personal experience, agricultural research tends to be very difficult to maintain with per-researcher operating funds of less than $100,000. As this sector becomes more active, public funds can be directed more efficiently to fewer crops of greater importance, in order to meet the demands of Kenya's rising population. It would also be advisable to improve research funding by diversifying its sources. Given the high potential for further development of plant breeding and biotechnology research, and the possibility of tapping into the rich expertise and ongoing programmes available in Kenya, such improvements could greatly benefit national and regional farmers.

As is often stated, biotechnology, especially GM research, relies on a firm foundation of classical breeding. While Kenya has such a foundation, the country has been slow to expand its capacity in biotechnology. The biotechnology budget in Kenya is only about a quarter the size of its plant breeding budget and funding decreased over the study period, both by discipline and by crop. If this trend continues, the amount of financial resources available to each scientist will also fall. The size of the Kenyan budget, and the fact that it does not seem entirely sufficient even at such levels, underscores the fact that plant breeding and biotechnology are expensive endeavors. Both research activities are labor intensive, field trial-dependent, and take many breeding cycles (time) before results are seen. In addition, it is important for breeders to constantly update their understanding of crops and strategies, keep their institute from becoming isolated, and help plan adoption strategies. The analysis revealed that the Kenyan plant breeding system emphasised line evaluation over the study period. In terms of future work, the Kenyan Agricultural Research Institute could take the lead in a strategic study of Kenyan crop priorities for breeding, rates of adoption, and the degree of international collaboration. Over the study period, varietal production appears to have matched expectations for the available funding. However, yield and other genetic improvements should be confirmed at the level of farmers' fields, and contributions to GDP value added should be assessed.

In addition, the interests and priorities of private sector breeding programmes should be contrasted with those of the public sector, with an aim toward exploring their complementarity and assessing which potential products could be quickly delivered. This would help create proper incentives for the private sector to continue investing in plant breeding and biotechnology, and to collaborate with the public sector in current initiatives. Such an analysis could lead to further explorations, such as possible public-private partnerships that may enable KARI and other public sector institutions to get their innovations into farmers' hands, by allowing the private sector to serve as a conduit for technology transfer. This approach would redirect KARI and the Kenyan public sector R&D institutions, encouraging them to bring their technologies up to a development level that demonstrates potential usefulness, thus supporting uptake by the private sector.

Philippines

Among the studied countries, the Philippines are subject to the highest population pressure that combined with the limited area available for agriculture, places the highest demands to increase agricultural productivity. This fact may partially explain the relatively high levels of biotechnology and plant breeding R&D funding by the public and private sector in the Philippines seen during the study period. In contrast to the other surveyed countries, researchers in the Philippines had per-researcher funding of a generous $196,532 in 2004. This indicates that the public sector places significant importance and trust on the research system, and believes that technology development is crucial to improving existing productivity. This is also reflected in the capabilities of the plant breeders and biotechnologists, and the complex nature of their programmes. One area of concern would be the need for drafting explicit policies to address the pending retirement of aging researchers, and the need to train and replace them with younger scientists. Policy makers should recognise that it is important to link older and newer researchers, in order to tap into the accumulated knowledge pool and ensure a smooth transition.

The observed funding per researcher in the Philippines was the highest among the four countries studied, and this investment supported the release of approximately 78 new varieties in 2004. The major emphasis of all investments was on rice, vegetables and fruit, and oilseed crops. The Philippines were involved in broad international collaborations, and had a solid in-country breeding programme at the Institute of Plant Breeding (IPB). Plant breeders were equally focused on germplasm evaluation and line development, with line evaluation receiving less attention, indicating that these researchers placed more emphasis on the development, rather than testing, of new varieties. This advanced capability is a major strength of the Philippines' R&D system.

In addition, unlike the other countries in this study, regulators and decision makers in the Philippines have approved the growth and use of commercially-provided GM maize for food and feed, and have invested public resources in the development of a number of GM crops. For this reason, the Philippines offer the best opportunity among the studied countries to explore the link between biotechnology and breeding in the public sector. In the future, this sort of study would be very useful for informing policy makers and helping guide resource allocations, and the results could be extended to the other countries in this study. The financial and human capacity for biotechnology is considerable in the Philippines, which had the highest annual budget for biotechnology among the studied countries. However, it is too soon to determine the absolute value of these efforts, as that can only really happen when public GM varieties are approved for commercialisation and made available to farmers.

Venezuela

Venezuela is a rich oil country with a very small agricultural sector. This country has the lowest population pressure of the four studied countries, and the smallest share for agriculture in the nation's GDP. Even though its plant breeding and biotechnology budget was the lowest among the four studied countries, Venezuela released among the most varieties from the fewest institutes (although the bulk of private sector releases were derived from varieties developed elsewhere). The largest focus for these releases was rice, maise, and legumes, followed by sorghum and millet, and vegetables and fruit. The human and financial resources invested by Venezuela were lower than those invested by the Philippines and Kenya. The main concerns identified in the present work are the need to assure congruence between investments in PGRFA use and conservation and crop priorities. Taking a long-term perspective, it may also be important to assess how Venezuela's petroleum-based economy will move into a post-petroleum economy, and how long-term investments in the common innovation infrastructure and cluster-specific areas of interest could support such a move. The process of shifting from a petroleum-based economy to a more diversified economy may be gradual, but it will take time to develop an innovation infrastructure and R&D targeted toward crop improvement innovation.

As noted above, Venezuelan institutes have typically tapped into varieties developed elsewhere, selecting and developing them for national needs. This approach has been complemented by internal efforts to select domestically-derived varieties. The established capacity to release varieties derived from spillovers may be complementary to the use of biotechnologies derived internally or externally. In the end, the quality of a given biotechnology product is heavily dependent on the quality of the parent germplasm, the gene product and the transformation process. Having a functional seed system is necessary (but not sufficient) to ensure the future viability and suitability of a biotechnology-supported agriculture. It will be necessary to assess the medium- to long-term viability of strategies aimed at utilising alternative pathways to address national capacities, versus continued use of spillovers and adaptation of innovations developed elsewhere. In practice, this should probably be based on a long-term plan by the Venezuelan government for the future transition to a less fossil fuel-dependent, diversified economy. Although agriculture in Venezuela has decreased in terms of its share of overall national GDP, it has not diminished in terms of the number of stakeholders and its value as a social valve. Thus, addressing the needs and constraints of the rural sector should be seen as a long-term investment with potentially enormous payoffs.

REFERENCES

Adato, M. and R. Meinzen-Dick. 2003. "Assessing the impact of agricultural research on poverty and livelihoods", *Quarterly Journal of International Agriculture* 42(2):149-167.

Conway, G., and V. Ruttan. 1999. *The Doubly Green Revolution: Food for All in the 21st Century*. London: Penguin Books.

Duvick, D.N. 1997. "Genetic rates of gain in hybrid maize yields during the past 40 years". *Maydica*, 22:187-196.

Evenson, R.E and D. Gollin. 2003. "Assessing the impact of the green revolution, 1960 to 2000". *Science* 300 (May): 758-762.

Furman, J. L., M. E. Porter, and S. Stern. 2002. "The determinants of national innovative capacity". *Research Policy* 31(6), 899–933.

Huang, J., C. Pray, and S. Rozelle. 2002. "Enhancing the crops to feed the poor". *Nature* 418 (August): 678-684.

Kishore, G. M. and C. Shewmaker. 1999. "Biotechnology: Enhancing human nutrition in developing and developed world". *PNAS* 93: 5968-5972.

United Nations Secretariat, Population Division of the Department of Economic and Social Affairs. 2004. *World Population Prospects: The 2004 Revision*. New York: United Nations. Available at http://esa.un.org/unpp.

3

Plant Breeding: Impact of Biotechnology

The year 2005 saw the tenth commercial planting season of genetically modified (GM) crops, which were first widely grown in 1996. In 2006, the billionth acre of GM crops was planted somewhere on Earth. These milestones provide an opportunity to critically assess the impact of this technology on global agriculture.

Although the first commercial GM crops were planted in 1994 (tomato), 1996 was the first year in which a significant area [1.66 million hectares (ha)] of crops were planted containing GM traits. Since then there has been a dramatic increase in plantings, and by 2005/06, the global planted area reached approximately 87.2 million ha. Almost all of the global GM crop area derives from soybean, maize (corn), cotton, and canola (Fig. 1). In 2005, GM soybean accounted for the largest share (62%) of total GM crop cultivation, followed by maize (22%), cotton (11%), and canola (5%). In terms of the share of total global plantings to these four crops accounted for by GM crops, GM traits accounted for a majority of soybean grown (59%) in 2005 (i.e., non-GM soybean accounted for 41% of global soybean acreage in 2005). For the other three main crops, the GM shares in 2005 of total crop production were 13% for maize, 27% for cotton, and 18% for canola.

The trend in plantings of GM crops (by crop) from 1996 to 2005 is shown in Figure 2. In terms of the type of biotechnology trait planted, Figure 1.3 shows that GM herbicide-tolerant soybean dominate, accounting for 58% of the total, followed by insect-resistant (largely Bt) maize and cotton with respective shares of 16% and 8%. In total, herbicide tolerant crops (GM HT) account for 76%, and insect resistant crops (GM IR) account for 24% of global plantings. Finally, looking at where biotech crops have been grown, the United States had the largest share of global GM crop plantings in 2005 (55%: 47.4 million ha), followed by Argentina (16.93 million ha: 19%of the global total). The other main countries planting GM crops in 2005 were Canada, Brazil, and China (Fig. 1.4).

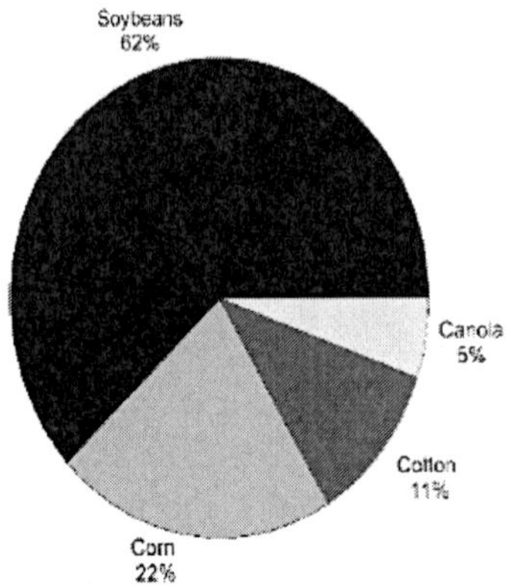

Figure 1: Global GM crop plantings in 2005 by crop (base area: 87.2 million ha).

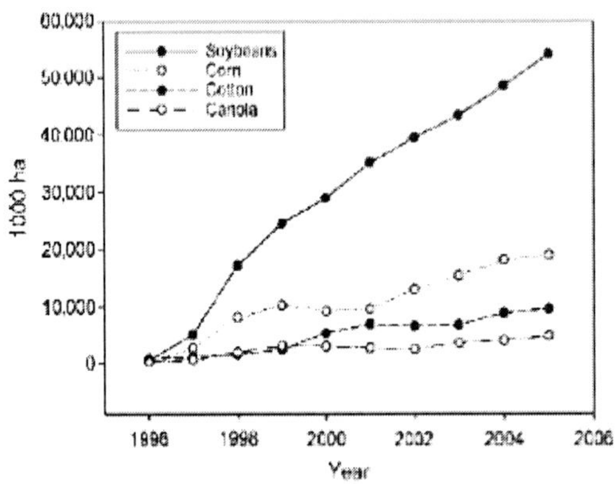

Figure 2: Global GM crop plantings by crop 1996–2005.

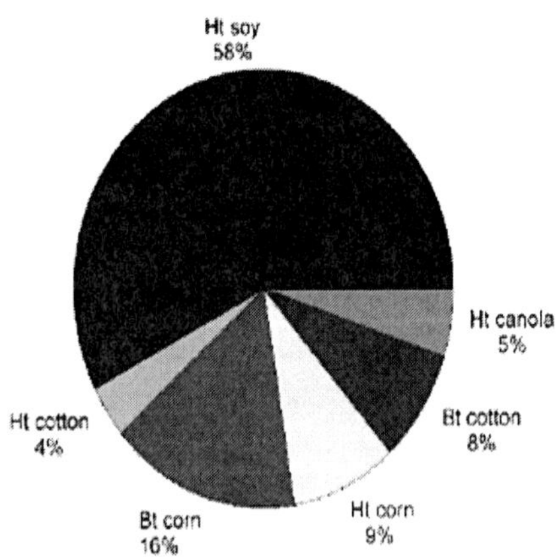

Figure 1.3. Global GM crop plantings by main trait and crop: 2005

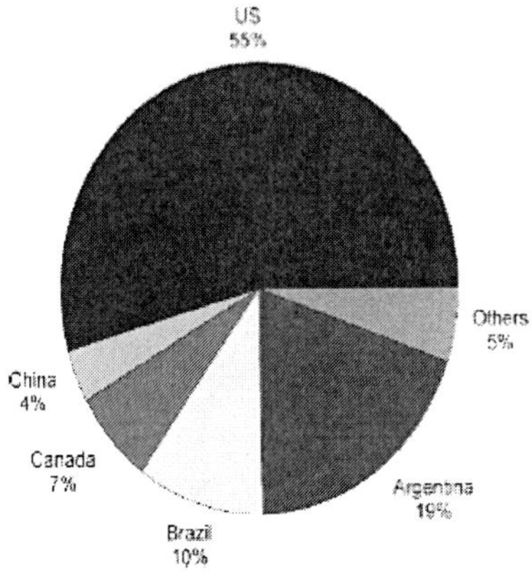

Figure 1.4. Global GM crop plantings 2005 by country.

WHY FARMERS USE BIOTECH CROPS

The primary driver of adoption among farmers (both large commercial and small-scale subsistence) has been the positive impact on farm income. The adoption of biotechnology has had a very positive impact on farm income derived mainly from a combination of enhanced productivity and efficiency gains (Table 1). In 2005, the direct global farm income benefit from GM crops was $5 billion. If the additional income stemming from second crop soybeans in Argentina is considered, this income gain rises to $5.6 billion. This is equivalent to having added between 3.6% and 4.0% to the value of global production of the four main crops of soybean, maize, canola, and cotton, a substantial impact. Since 1996, worldwide farm incomes have increased by $24.2 billion or $27 billion inclusive of second-crop soybean gains in Argentina directly because of the adoption of GM crop technology.

Table 1: Global Farm Income Benefits from Growing GM Crops 1996–2005 (million US $)

Trait and Crop	Increase in Farm Income, 2005	Increase in Farm Income, 1996–2005	Farm Income Benefit in 2005 as % of Total Value of Production of These Crops in GM-Adopting Countries	Farm Income Benefit in 2005 as % of Total Value of Global Production of These Crops
GM herbicidetolerant	2281	11,686	5.72	4.86
Soybean	(2842)	(14,417)	(7.1)	(6.05)
GM herbicidetolerant maize	212	795	0.82	0.39
GM herbicidetolerant cotton	166	927	1.16	0.64
GM herbicidetolerant canola	195	893	9.45	1.86
GM insect-resistant maize	416	2,367	1.57	0.77
GM insect-resistant cotton	1,732	7,510	12.1	6.68
Others	25	66	–	–
Totals	5027	24,244	6.0	3.6
	(5588)	(26,975)	(6.7)	(4.0)

The largest gains in farm income have arisen in the soybean sector, largely from cost savings, where the $2.84 billion additional income generated by GM HT soybean in 2005 has been equivalent to adding 7.1% to the value of the crop in the GM-growing countries, or adding the equivalent of 6.05% to the $47 billion value of the global soybean crop in 2005. These economic benefits should, however, be placed within the context of a significant increase in the level of soybean production in the main GM-adopting countries. Since 1996, the soybean area and production in the leading soybean producing countries of the United States, Brazil, and Argentina increased by 58% and 65%, respectively. Substantial gains have also arisen in the cotton sector through a combination of higher yields and lower costs. In 2005, cotton farm income levels in the GM-adopting countries increased by $1.9 billion and since 1996, the sector has benefited from an additional $8.44 billion. The 2005 income gains are equivalent to adding 13.3% to the value of the cotton crop in these countries, or 7.3% to the $26 billion value of total global cotton production. This is a substantial increase in value-added terms for two new cotton seed technologies.

Significant increases to farm incomes have also resulted in the maize and canola sectors. The combination of GM IR and GM HT technology in maize has boosted farm incomes by over $3.1 billion since 1996. An additional $893 million has been generated in the North American canola sector. Overall, the economic gains derived from planting GM crops have been of two main types: (1) increased yields (associated mostly with GM insect-resistant technology) and (2) reduced costs of production derived from less expenditure on crop protection (insecticides and herbicides) products and fuel. Table 2 summarises farm income impacts in key GM-adopting countries highlighting the important direct farm income benefit arising from growing GM HT soybeans in Argentina, GM IR cotton in China, and a range of GM cultivars in the United States.

Table 2: GM Crop Farm Income Benefits during 1996–2005 in Selected Countries (million US $)

Country	GM HT Soybean	GM HT Maize	GM HT Cotton	GM HT Canola	GM IR Maize	GM IR Cotton	Total
USA	7570	771	919	101	1957	1627	12,945
Argentina	5197	0.2	4.0	N/A	159	29	5389.2
Brazil	1367	–	–	–	–	–	1367
Paraguay	132	–	–	–	–	–	132
Canada	69	24	–	792	145	–	1031
South Africa	2.2	0.3	0.2	–	59	14	75.7
China	–	–	–	–	–	5168	5168
India	–	–	–	–	–	463	463
Australia	–	–	4.1	–	–	150	154.1
Mexico	–	–	–	–	–	55	55
Philippines	–	–	–	–	8	–	8
Spain	–	–	–	–	28	–	28

In terms of the division of the economic benefits, it is interesting to note that farmers in developing countries derived the majority of the farm income benefits in 2005 (55%) relative to farmers in developed countries (Table 3). The vast majority of these income gains for developing country farmers have been from GM IR cotton and GM HT soybean.

Table 3: GM Crop Farm Income Benefits, 2005: Developing Versus Developed Countries (US $)

Crop	Developed	Developing	% Developed	% Developing
GM HT soybean	1183	1658	41.6	58.4
GM IR maize	364	53	86.5	13.5
GM HT maize	212	0.3	99.9	0.1
GM IR cotton	354	1378	20.4	79.6
GM HT cotton	163	3	98.4	1.6
GM HT canola	195	0	100	0
GM VR papaya and squash	25	0	100	0
Totals	2496	3092	45	55

Examination of the cost farmers pay for accessing GM technology relative to the total gains derived shows (Table 4) that across the four main GM crops, the total cost was equal to about 26% of the total farm income gains. For farmers in developing countries the total cost is equal to about 13% of total farm income gains, while for farmers in developed countries the cost is about 38% of the total farm income gain. Although circumstances vary among countries, the higher share of total gains derived by farmers in developing countries relative to farmers in developed countries reflects factors such as weaker provision and enforcement of intellectual property rights.

Table 4: Cost of Accessing GM Technologya (in %) Relative to Total Farm Income Benefits, 2005

Crop	All Farmers	Developed Countries	Developing Countries
GM HT soybean	21	32	10
GM IR maize	44	43	48
GM HT maize	38	38	81
GM IR cotton	21	41	13
GM HT cotton	44	43	65
GM HT canola	47	47	—
Totals	26	38	13

In addition to the tangible and quantifiable impacts on farm profitability presented above, there are other important, more intangible (difficult to quantify) impacts of an economic nature. Many studies on the impact of GM crops have identified the factors listed below as being important influences for adoption of the technology.

Herbicide-Tolerant Crops

1. This method provides increased management flexibility due to a combination of the ease of use associated with broad-spectrum, postemergent herbicides like glyphosate (often referred to by its more commonly known brand name of Roundup) and the increased/longer time window for spraying.
2. In a conventional crop, postemergentweed control relies on herbicide applications before theweeds and crop arewell established. As a result, the cropmay suffer "knockback" to its growth from the effects of the herbicide. In the GM HT crop, this problem is avoided because the crop is tolerant to the herbicide and spraying can occur at a later stage when the crop is better able to withstand any possible knockback effects.
3. This method facilitates the adoption of conservation or no-tillage systems. This provides for additional cost savings such as reduced labor and fuel costs associated with plowing.
4. Improved weed control has contributed to reduced harvesting costs—cleaner crops have resulted in reduced times for harvesting. It has also improved harvest quality and led to higher levels of quality price bonuses in some regions (e.g., Romania).
5. Potential damage caused by soil-incorporated residual herbicides in follow-on crops has been eliminated.

Insect-Resistant Crops

1. For production risk management/insurance purposes, this method eliminates the risk of significant pest damage.
2. A "convenience" benefit is derived because less time is spent walking through the crop fields to survey insects and insect damage and/or apply insecticides.
3. Savings in energy use are realised—associated mainly with less frequent aerial spraying.
4. There are savings inmachinery use (for spraying and possibly reduced harvesting times).
5. The quality of Bt maize is perceived as superior to that of non-Bt maize because the level of fungal (Fusarium) damage, which leads to mycotoxin presence in plant tissues, is lower with Bt maize. As such, there is an increasing body of evidence that Fusarium infection levels and mycotoxin levels in GM insect resistant maize are significantly (5–10-fold) lower than those found in conventional (nonbiotech) crops. This lower mycotoxin contamination in turn leads to a safer food or feed product for consumption.

6. There Health and safety for farmers and farmworkers is improved (handling and use of pesticides is reduced).
7. The growing season is shorter (e.g., for some cotton growers in India), which allows some farmers to plant a second crop in the same season (notably maize in India). Also some Indian cotton growers have reported commensurate benefits for beekeepers as fewer bees are now lost to insecticide spraying.

IMPACTS OF ADOPTION OF PLANT BIOTECHNOLOGY

The two key aspects of environmental impact of biotech crops examined below are decreased insecticide and herbicide use, and the impact on carbon emissions and soil conservation.

Environmental Impacts

Usually, changes in pesticide use with GM crops have traditionally been presented in terms of the volume (quantity) of pesticide applied. While comparisons of total pesticide volume used in GM and non-GM crop production systems can be a useful indicator of environmental impacts, it is an imperfect measure because it does not account for differences in the specific pest control programmes used in GM and non-GM cropping systems. For example, different specific chemical products used in GM versus conventional crop systems, differences in the rate of pesticides used for efficacy, and differences in the environmental characteristics (mobility, persistence, etc.) are masked in general comparisons of total pesticide volumes used.

To provide a more robust measurement of the environmental impact of GM crops, the analysis presented below includes an assessment of both pesticide active-ingredient use and the specific pesticides used via an indicator known as the environmental impact quotient (EIQ). This universal indicator, developed by Kovach et al. 1992 and updated annually, effectively integrates the various environmental impacts of individual pesticides into a single field value per hectare. This index provides a more balanced assessment of the impact of GM crops on the environment as it draws on all of the key toxicity and environmental exposure data related to individual products, as applicable to impacts on farmworkers, consumers, and ecology, and provides a consistent and comprehensive measure of environmental impact. Readers should, however, note that the EIQ is an indicator only and therefore does not account for all environmental issues and impacts. The EIQ value is multiplied by the amount of pesticide active ingredient (AI) used per hectare to produce a field EIQ value. For example, the EIQ rating for glyphosate is 15.3. By using this rating multiplied by the amount of glyphosate used per hectare (e.g., a hypothetical example of 1.1 kg applied per hectare), the field EIQ value for glyphosate would be equivalent to 16.83/ha. In comparison, the field EIQ/ha value for a commonly used herbicide on corn crops (atrazine) is 22.9/ha.

The EIQ indicator is therefore used for comparison of the field EIQ/ha values for conventional versus GM crop production systems, with the total environmental impact or load of each system, a direct function of respective field EIQ/ha values, and the area planted to each type of production (GM vs. non-GM). The EIQ methodology is used below to calculate and compare typical EIQ values for conventional and GM crops and then aggregate these values to a national level. The level of pesticide use in the respective areas planted for conventional and GM crops in each year was compared with the level of pesticide use that probably would otherwise have occurred if the whole crop, in each year, had been produced using conventional technology (based on the knowledge of crop advisers). This approach addresses gaps in the availability of herbicide or insecticide usage data in most countries and differentiates between GM and conventional crops. Additionally, it allows for comparisons between GM and non-GM cropping systems when GM accounts for a large proportion of the total crop planted area. For example, in the case of soybean in several countries, GM represents over 60% of the total soybean crop planted area. It is not reasonable to compare the production practices of these two groups as the remaining non-GM adopters might be farmers in a region characterised by below-average weed or pest pressures or with a tradition of less intensive production systems, and hence, below-average pesticide use.

GM crops have contributed to a significant reduction in the global environmental impact of production agriculture (Table 5). Since 1996, the use of pesticides was reduced by 224 million kg of active ingredient, constituting a 6.9% reduction, and the overall environmental impact associated with pesticide use on these crops was reduced by 15.3%. In absolute terms, the largest environmental gain has been associated with the adoption of GM HT soybean and reflects the large share of global soybean plantings accounted for by GM soybean. The volume of herbicide use in GM soybean decreased by 51 million kg since 1996, a 4.1% reduction, and the overall environmental impact decreased by 20%. It should be noted that in some countries, such as in Argentina and Brazil, the adoption of GM HT soybean has coincided with increases in the volume of herbicides used relative to historic levels. This net increase largely reflects the facilitating role of the GM HT technology in accelerating and maintaining the switch away from conventional tillage to no/ low-tillage production systems, along with their inherent environmental benefits. This net increase in the volume of herbicides used should, therefore, be placed in the context of the reduced GHG emissions arising from this production system change and the general dynamics of agricultural production system changes. Major environmental gains have also been derived from the adoption of GM insectresistant (IR) cotton. These gains were the largest of any crop on a per hectare basis. Since 1996, farmers have used 95.5 million kg less insecticide in GM IR cotton crops (a 19.4% reduction), and reduced the environmental impact by 24.3%. Important environmental gains have also arisen in the maize and canola sectors.

Table 5: Impact of Changes in Use of Herbicides and Insecticides from Growing GM Crops Globally, 1996–2005

Trait	Change in Volume of Active Ingredient Used (million kg)	Change in Field EIQ Impact	% Change in AI use in GM-Growing Countries	% Change in Environmental Impact in GM-Growing Countries
GM herbicidetolerant soybean	–51.4	–4,865	–4.1	–20.0
GM herbicidetolerant maize	–36.5	–845	–3.4	–4.0
GM herbicidetolerant cotton	–28.6	–1,166	–15.1	–22.7
GM herbicidetolerant canola	–6.3	–310	–11.1	–22.6
GM insectresistant maize	–7.0	–403	–4.1	–4.6
GM insectresistant cotton	–94.5	–4,670	–19.4	–24.3
Totals	–224.3	–12,259	–6.9	–15.3

In the maize sector, pesticide use decreased by 43 million kg and the environmental impact decreased because of reduced insecticide use (4.6%) and a switch to more environmentally benign herbicides (4%). In the canola sector, farmers reduced herbicide use by 6.3 million kg (an 11% reduction) and the environmental impact has fallen by 23% because of a switch to more environmentally benign herbicides. The impact of changes in insecticide and herbicide use at the country level (for the main GM-adopting countries) is summarised in Table 6.

Table 6: Reduction in "Environmental Impact" from Changes in Pesticide Use Associated with GM Crop Adoption by Country, 1996–2005, Selected Countries

Country	GM HT Soybean	GM HT Maize	GM HT Cotton	GM HT Canola	GM IR Maize	GM IR Cotton
USA	29	4	24	38	5	23
Argentina	21	NDA	NDA	–	0	4
Brazil	6	–	–	–	–	–
Paraguay	13	–	–	–	–	–
Canada	9	5	–	22	NDA	–
South Africa	7	0.44	6	N/A	2	NDA
China	–	–	–	–	–	28
India			–	–	–	3
Australia		–	4	–	–	22
Mexico	–	–	–	–	–	NDA
Spain	–	–	–	–	30	–

In terms of the division of the environmental benefits associated with less insecticide and herbicide use for farmers in developing countries relative to farmers in developed countries, Table 1.7 shows that in 2005, the majority of the environmental benefits associated with lower insecticide and herbicide use have been for developing-country farmers. The vast majority of these environmental gains have been from the use of GM IR cotton and GM HT soybeans.

Table 7: GM Crop Environmental Benefits from Lower Insecticide and Herbicide Use in 2005: Developing versus Developed Countries

Crop	Percent of Total Reduction in Ela	
	Developed Countries	Developing Countries
GM HT soybean	53	47
GM IR maize	92	8
GM HT maize	99	1
GM IR cotton	15	85
GM HT cotton	99	1
GM HT canola	100	0
Totals	46	54

Impact on Greenhouse Gas (GHG) Emissions

Reductions in the level of GHG emissions from GM crops are from two principal sources: First, GM crops contribute to a reduction in fuel use from less frequent herbicide or insecticide applications and a reduction in the energy use in soil cultivation. For example, Lazarus and Selley estimated that one pesticide spray application uses 1.045 liters (L) of fuel, which is equivalent to 2.87 kg/ha of carbon dioxide emissions. In this analysis we used the conservative assumption that only GM IR crops reduced spray applications and ultimately GHG emissions. In addition to the reduction in the number of herbicide applications there has been a shift from conventional tillage to no/reduced tillage. This has had a marked effect on tractor fuel consumption because energy-intensive cultivation methods have been replaced with no/ reduced tillage and herbicide-based weed control systems. The GM HT crop where this is most evident is GM HT soybean. Here, adoption of the technology has made an important contribution to facilitating the adoption of reduced/no-tillage (NT) farming. Before the introduction of GM HT soybean cultivars, NT systems were practiced by some farmers using a number of herbicides and with varying degrees of success. The opportunity for growers to control weeds with a nonresidual foliar herbicide as a "burndown" preseeding treatment, followed by a postemergent treatment when the soybean crop became established, has made the NT system more reliable, technically viable, and commercially attractive. These technical advantages, combined with the cost advantages, have contributed to the rapid adoption of GM HT cultivars and the near-doubling of the NT soybean area in the United States (and also a 5-fold increase in Argentina). In both countries, GM HT soybean crops are estimated to account for 95% of the NT soybean crop area. Substantial growth in NT production systems has also occurred in Canada, where the NT canola area increased from 0.8 to 2.6 million ha (equal to about half of the total canola area) between 1996 and

2005 (95% of the NT canola area is planted with GM HT cultivars). Similarly, the area planted to NT in the US cotton crop increased from 0.2 to 1 million ha over the same period (86% of which is planted to GM HT cultivars). The increase in the NT cotton area has been substantial from a base of 200,000 ha to over 1.0 million ha between 1996 and 2005. The fuel savings resulting from changes in tillage systems are drawn from estimates from studies by Jasa and CTIC. The adoption of NT farming systems is estimated to reduce cultivation fuel usage by 32.52 L/ha compared with traditional conventional tillage and 14.7 L/ha compared with (the average of) reduced tillage cultivation. In turn, this results in reductions in CO emissions of 89.44 and 40.43 kg/ha, respectively.

Second, The use of reduced/no-tillage farming systems that utilise less plowing increase the amount of organic carbon in the form of crop residue that is stored or sequestered in the soil. This carbon sequestration reduces carbon dioxide emissions to the environment. Rates of carbon sequestration have been calculated for cropping systems using normal tillage and reduced tillage, and these were incorporated in our analysis on how GMcrop adoption has significantly facilitated the increase in carbon sequestration, ultimately reducing the release of CO_2 into the atmosphere. Of course, the amount of carbon sequestered varies by soil type, cropping system, and ecoregion. In North America, the International Panel on Climate Change estimates that the conversion from conventional tillage to no-tillage systems stores between 50 and 1300 kg C/ha annually (average 300 kg C/ha per year).

Table 8: Impact of GM Crops on Carbon Sequestration Impact in 2005; Car Equivalents

Crop/Trait/Country	Permanent CO_2 Savings from Reduced Fuel Use (million kg CO_2)	Average Family Car Equivalents Removed from Road per Year from Permanent Fuel Savings	Potential Additional Soil Carbon Sequestration Savings GM-Growing (million kg CO_2)	Average Family Car Equivalents Removed from Road per Year from Potential Additional Soil Carbon Sequestration
US: GM HT soybean	176	78,222	2,195	975,556
Argentina: GM HT soybean	546	242,667	4,340	1,928,889
Other countries: GM HT soybeans	55	24,444	435	193,333
Canada: GM HT canola	117	52,000	1,083	481,520
Global GM IR cotton	68	30,222	0	0
Totals	962	427,556	8,053	3,579,298

In the analysis presented below, a conservative savings of 300 kg C/ha per annum was applied to all no-tillage agriculture and 100 kg C/ha^{-1} year^{-1} was applied to reduced-tillage agriculture. Where some countries aggregate their no/reduced-tillage data, the reduced-tillage saving value of 100 kg C/ha–1 year–1 was used. One kilogram of carbon sequestered is equivalent to 3.67 kg of carbon dioxide. These assumptions were applied to the reduced pesticide spray applications data on GM IR crops, derived from the farm income literature review, and the GM HT crop areas using no/reduced tillage (limited to the GM HT soybean crops in North and South America and GM HT canola crop in Canada). Table 1.8 summarises the impact on GHG emissions associated with the planting of GM crops between 1996 and 2005. In 2005, the permanent CO2 savings from reduced fuel use associated with GM crops was 0.962 billion kg. This is equivalent to removing 430,000 cars from the road for a year. The additional soil carbon sequestration gains resulting from reduced tillage with GM crops accounted for a reduction in 8.05 billion kg of CO2 emissions in 2005. This is equivalent to removing nearly 3.6 million cars from the roads per year. In total, the carbon savings from reduced fuel use and soil carbon sequestration in 2005 were equal to removing 4 million cars from the road (equal to 17% of all registered cars in the UK).

REFERENCES

Brookes G, Barfoot P (2007): GM crops: The first ten years—global socio-economic and environmental impacts. *AgbioForum* 9:1–13.

Conservation Tillage and Plant Biotechnology (CTIC) (2002): *How New Technologies Can Improve the Environment by Reducing the Need to Plough* (http://www.ctic.purdue.edu/CTIC/Biotech. html).

James C (2006): *Global Status of Transgenic Crops, Various Global Review Briefs from 1996 to 2006*. International Service for the Acquisition of Agri-Biotech Applications (ISAAA).

Jasa P (2002): *Conservation Tillage Systems*, Extension Engineer, Univ Nebraska.

Kovach J, Petzoldt C, Degni J, Tette J (1992): *A Method to Measure the Environmental Impact of Pesticides*. New York's Food and Life Sciences Bulletin, NYS Agricultural Experiment Station, Cornell Univ, Geneva, NY, p 139, 8 pp annually updated (http://www.nysipm.cornell.edu/ publications/EIQ.html).

Lazarus W, Selley R (2005): *Farm Machinery Economic Cost Estimates for 2005*, Univ Minnesota Extension Service.

Trigo et al. (2002): *Genetically Modified Crops in Argentina Agriculture: An Opened Story*. Libros del Zorzal Buenos Aires, Argentina.

4

Organic Plant Breeding

The terms: "organic plant breeding", "plant breeding for the organic agriculture" and "organic propagation" describe different topic areas. They need to be exactly defined and must not be used as synonyms in discussion. It is also important to ensure that political and commercial agendas cannot influence the development of sensible breeding targets, strategies and methodologies, which are focused on the production of crop varieties adapted for organic production systems.

Organic agriculture is regulated by standards at different levels. In the EU there is the Council Regulation on organic agriculture, which also includes livestock standards since 1999. Some countries have published additional governmental standards. Furthermore there are organic farming standards produced by organic farming sector bodies (e.g. Naturland, Bioland, Soil Association etc.) which prescribe an even higher level of farming standards in many countries. All these standards do not include specific regulations about plant breeding and seed production. EU Council Regulation prescribes the use of seed which have been propagated one year under organic management . This regulation will come into effect on the 1.1.2004 , until then organically produced seed only have to be used if they are available. All organic farming standards insist on the exclusion of genetic modified organism. As there is an increasing interest in the selection and/or plant breeding of crop varieties which are suitable for sustainable and organic production systems, it is expected that organic standards regulating seed production and crop breeding will be developed over the next 2-3 years. Until recently discussions about "organic plant breeding standards" mainly involved bio-dynamic farmers and their sector body representives However, they often described their "biodynamic" approach under the theme „plant breeding for an organic agriculture". In some publications topics such as the necessity of organic propagation from 2003, the impact of organic seed propagation methods on with organic breeding and/or the prohibition on genetic engeneering were combined with a the call for the prohibition on the use of hybrid seed. Before starting a discussion on „"breeding for organic and sustainable production" it is therefore essential to separate and define these different topics.

ORGANIC BREEDING AND PROPAGATION

The main principle of organic agriculture which is to establish a holistic approach to food production will inevitably require a similar approach for crop breeding . The discussion about the definition of "organic plant breeding" and what the criteria are to separate conventional from organic plant breeding is not finished yet. Based on the current discussions (which took place mainly among biodynamic farming groups and organisations) the following hirarchy for different breeding tools/methods has been developed:

(a) are prefered

(b) are allowed, but in a period of approx. 5 years to check (in other countries: provisionally certified)

(c) are not suitable

COMMON MISCONCEPTIONS ABOUT THE CHOICE OF VARIETIES IN ORGANIC FARMING

When considering the seed/variety requirements of the organic farming industry one has to first discuss a range of commonly made statements:

"The conventional varieties are unsuitable for the ecological agriculture"

The importance of selecting and/or breeding of varieties for characteristics important for organic farming (e.g. resistance against pests and diseases, ability to compete with weeds, nutrient use efficiency) is undisputed. Also because many of the inputs used in conventional farming (e.g. water soluble N and P fertilizers and KCL) are not permitted, the priorities for variety selection may be different (e.g. competitiveness against weeds is unimportant in conventional farming when effective selective herbicides are available), but is a main selection criterion in organic systems where herbicides are not permitted and mechanical, flame and/or hand weeding represent the main variable cost factor.

However, many selection criteria (e.g. resistance to many pests and diseases and abiotic stress factors) have similar importance in both organic and conventional systems and general statement such as: *"The conventional varieties are unsuitable for the ecological agriculture"* are clearly incorrect. At present varieties suitable for organic production systems are available/certified. for nearly all crop species (with the exception of barley varieties of poor soils). However, for some other crops breeders have more recently selected for specific characteristics, which may not be not suitable for organic production systems (short straw in some cereals). It is therefore currently difficult to assess, whether there is potential for special breeding and selection programmes to develop varieties specifically for organic farming systems. But it is sure that more suitable varieties will be developed step by step.

"The organic agriculture wants to cultivate old varieties"

This is very much amisconception and very much the opposite is true. Ecological agriculture wants to share in the progress of breeding, because organic farmers also have to pay attention to productivity and profibiltiy.

However, organic farmers may favour varieties that have resulted from "conventional" breeding programmes some 10 to 30 years ago. This may be because of:

1. more recent varieties lack specific charcteristics (e.g. better organoleptic characteristics; in cereals longer straw, which is valued because it increases the competitiveness against weeds and provides more bedding material for livestock)
2. organic farmers have to consider the resource requirements of crops within the framework of their entire production system (e.g. a higher yield and/or a higher nutrient uptake efficiency of one crop may be detrimental to successive crops, and benefits of specific varieties always habe to be calculated within the whole rotation).
3. there are insufficient research data on the performance of newer varieties under organic management practices,
4. there is a lack of non-treated and/or organically produced seed of older varieties

However a sufficient profibility of the cultivation of old varieties must be made possible by additional govermentional money (like already intended in some state programmes).

"The number of the varieties for the organic agriculture decreases."

Indeed varieties which were highly suitable for organic production systems have been removed from national variety-lists again and again and were not always replaced with equivalents (eg. this happened with baking quality-wheat or high organoleptic quality ware potato varieties). Nevertheless the choice of varieties available to organic farmers is higher now than it was 10 years agao, which is mainly due better access to interesting varieties from other EU-countries. However, in the future, the choice of verities available to organic farmers may become more restricted, if genetically modified crop varieties can be grown on a commercial scale in the EU.

"Eecological agriculture needs regionally adapted varieties".

Adaptation of crops to regional soil, climate and production systems is an important variety characteristic. However, agronomic benefits of selecting/breeding local adaptation varies and the definition of a "region" varies between crop species. Regions can be divided into different groups respecting their location-factors. For example there are not too many regions in Germany: Dry and moist regions, good and bad soils, mild winter with longer time for growing and shorter time of vegetation.

Breeding companies will always have to consider the additional benefit but also costs associated with developing a wide range of varieties which are adapted to very specific conditions and the seed production economies associated with varieties which can be used for a wide range of locations. Because of the need for profitability for the relatively smaller (organic) segment of seed market, breeding programmes for organic agriculture should initially focus on the development of varitieties suitable for many locations.

"Organic farmers are looking for varieties to propagate and save them on their farms"

You need a conservation breeding to keep a variety in its quality. This can be done by farmers on their farms theoretically. But most of the farmers are not ecucated to do that and also don't want to do that.

"The best varieties for conventional farmers are the best varieties for the organic farmers, too"

This statement is also not generally true. There are traits associated with conventional varieties which are unsuitable for organic production systems and certain traits required in organc farming systems are not present in recently developed "conventional" varieties. This includes

1. Ability to compete with weeds. The ability to supress weeds is an important trait in both cereals and non-cereals (eg peas).
2. Longer straw. Cereal varieties with longer straw often have got more mass of roots, so they have got a higher ability for adaption of nutrients (important under conditions of lower concentration of nutrients).
3. However, for other traits which address problems found in both organic and conventional production modern conventional varieties may be the most suitable. In cereals this for example applies resistances to diseases which have similar importance in both organic and conventional farming such as necrothrophic/ opportunistic pathogens (eg Helmintosporium, Rynchosproium), but also some biothrophic pathogens (e.g. rusts)

VARIETY TRAITS FOR ORGANIC AGRICULTURE

Many variety traits in varieties bred for conventional agriculture are also required in organic farming, but the priority given to individual traits is often different. A high level of disease resistance is necessary especially with respect to Fusarium, Helmintosporium, Septoria, Claviceps and viruses. An intermediate level of resistance is sufficient for many biothrophic fungal diseases eg mildew, which decrease in importance in organic management systems. It is important to identify the exact balance of characterstics required and not to miss out specific traits altogether. Important regional requirements, eg resistance to rust, must be considerate just like in conventional agriculture.

Specific variety traits required for "organic" cereal varieties:

1. resistance or tolerance against bunt (tilletia tritici) and other seed borne diseases. If we want to establish organic seed production and conservation breeding systems, problems with seed borne diseases are likely to increase. For the near future we may be able to develop organically acceptable seed treatment methods, but in the longer term there is pressure to replace treatments with resistence breeding.
2. ability to supress weeds, longer straw
3. stability of yield and quality under different conditions (nutrient use efficiency, resistance to abiotic stress factors)
4. high quality and stability of quality characteristics under extensive conditions

Variety traits for leguminoses (combined varieties)

1. The existing diseases of leguminoses are a higher problem because of the higher frequency in organic agriculture

Variety traits for potatoes

1. Phytophtora infestans (late blight). Phytophtora causes variable yields. Considering the pending prohibition on the use of copper fungicides and the increased aggressiveness of the *Phytophthora* pathovar spectrum present in the EU this is the most important problem for the increasing organic potato market.. Although there are some new varieties with higher resistance levels, many have low organoleptic (taste) quality, poor visual appearance and/or limited storability.
2. Rhizoctonia

Variety traits for sugar beats, maize, sunflower and oil seed rape

No results of organic variety trails available.. The suitability of the currently available varieties can therefore not be assessed.

1. *Sugar beet:* Because of the great demand for organic sugar (especially from food processors) sugar beet would be a commercially interesting crop for farmers in areas where casual labout is available. Rapid growth during the early growth stages is thought to be important in areas with high abundance of young plant pests. However, there are no processors available for organic sugar beet in several EU-countries This is disappointing since sugar beats is an excellent break crop (providing both agronomic and economic benefits) for organic arable rotations especially in regions with fertile soils.
2. *Maize:* There are already interesting breeding programmes focusing on the development of varieties for low-input conditions. Again the ability to establish/

grow rapidly during the early growth stages is thought to be a desired characteristic for „organic varieties".

3. *Sunflower:*There is currently no difference between the characteristics desired for organic and conventional production systems. High straw strength and early maturing being important selection criteria as well as high yield.

4. *Oil seed rape:* It would be very desirable to have an alternative oil/animal feed protein crop to sunflower available in Germany. In Germany organic rape is attacked by several animal pests with resulting in yields of under 10 dt per ha. Another problem with winter rape are weeds and the high fertility demand. It is therefore necessary to carry out fundamental research in the development of organic rape varieties and agronomic strategies use to produce oil seed rape crops in Germany.

SHORT TERM AND MIDDLE TERM TASKS

If goverments want to increase number of organic farms, there has to be an increase in research underpinning organic agriculture.We propose the following objectives/tasks for the next 5 years:

1. Evaluation of existing varieties/breeding lines (both older and newer ones) for characteristics/traits required in organic farming systems

2. Organic variety trials for varieties of crops which were not yet widely tested for their performance under organic management practices: maize, sunflower, rape, gras, clover. In some countries also cereals, peas, beans, soja, lupins.

3. Research and developement of seed production, processing and conservation breeding for the organic agricultureSelection of breeding material under organic conditions. Evaluation and development of variety mixtures

4. Evaluation of breeding methods for organic breeding. Developing alternatives to not suitable breeding methods.

LONGER TERM PERSPECTIVES AND OUTLOOK

Longer term perspectives will depend on developments in arange of areas and is currently very difficult to predict:

1. If genetic modified varieties will be grown on a large scale within the EU – will there be breeding of non-genetic modified varieties? Can organic agriculture only react and conserve old varieties? Will the increase in organic agriculture make it commercially interesting to breed specifically for organic agriculture?

2. How can we create a balance between a breeding for the organic agriculture and an organic breeding, so that the organic farmers will harvest the greatest benefit? Organic agriculture cannot (and doesn't have to) work on all the crops (especially

not on the less important), but nevertheless the share of organcially bred varieties should increase.

ORGANIC GREENHOUSE VEGETABLE BREEDING

Although several Extension bulletins are available on greenhouse vegetable production, few of these concentrate on organic production methods. This publication presents an overview of greenhouse production systems and profiles several farmers raising organic vegetables in greenhouses. It is hoped this will give new growers ideas of how to set up their systems, and provide more experienced farmers with examples of alternative methods of production. The term greenhouse means different things to different people. A greenhouse used to be a structure formed of glass, with a heating (and usually cooling) system that was used year-round, but especially in winter. Then came houses built of thermoplastic (Plexiglas and others), followed by Quonsets covered with plastic, which may or may not be heated, have one or two layers, and be used year-round or for only a few months every year. The type of greenhouse you have will largely be determined by your crop and your capital and, to a lesser extent, by your management intensity and your market strategy.

The U.S. greenhouse vegetable industry is a mixture of small, family-run operations in the 2,500 to 10,000 square foot range and a small number of large, multi-acre facilities 10 acres or more in size. The larger greenhouses often use waste heat from a power plant or other source of cogeneration. Current U.S. production estimates are somewhere around 800 acres. In comparison, Mexico has about 450 acres, Canada has about 1,600 acres, and Holland has over 11,000 acres. In the latter part of the 1990s, Canadian greenhouse vegetable production grew at a rate of 20% a year. How has Canada been able to generate this huge growth? "Significant new greenhouse vegetable production technology that was transferred to commercial producers has been primarily responsible for dramatic yield increases over the last 7-8 years, estimated at 100-120% for tomatoes and 70-80% for cucumbers." Canada is strongly supporting its greenhouse growers, both with research and with investment dollars. Their research facilities at Harrow are recognised as topnotch the world over. Although most Canadian greenhouse vegetables are not produced organically, there has been an emphasis of late to use IPM strategies, rather than pesticides, to accommodate the growing market of consumers who want pesticide-free produce. Most of the organic produce imported into the U.S. is now coming from Mexico. The leading states in greenhouse vegetable production are California, Florida, Colorado, Arizona, Ohio, Texas, and Pennsylvania–all with over one million square feet of production each. (2) This is where the centers of production exist, both as a source of competition as well as a source of technical assistance. Small-scale growers who plan on direct market sales may want to think twice before getting into this business in a region that is already saturated.

With so much competition from Canada, Mexico, and overseas, how can small farmers realise a profit raising greenhouse vegetables? One issue of increasing importance to consumers is vegetables grown with minimum pesticides. The public has also become educated on the values of locally grown produce: it's fresher, it tastes better, and it may even be less expensive, since there are fewer shipping costs involved. Also, money paid to a local farmer is re-invested in the local community and helps to keep that economy strong.

Year-round production is key to maintaining the greenhouse's profitability. However, this does not necessarily mean that growers should be producing the same crop year-round. (Winter tomatoes bring more money than do summer ones.) Another option would be to raise a crop other than vegetables, like bedding plants for early spring sales or poinsettias for Christmas. The grower may decide that the most cost-efficient way to use his or her greenhouse during the summer is to shut it up to solarise the soil and "cook" insects (and their eggs) that are present.

Small growers must find niche markets. It is pointless to try to compete with mass merchandisers like Wal-Mart, because the small grower will always lose. What are some niche markets for organic greenhouse vegetable producers? Some of the general niches have already been mentioned: consumers are looking for organic, locally grown, early-season produce. Whatever the niche market, it is important for growers to realise that the nature of niche markets is for them to disappear after a while. Oversupply or lowered demand will create lower prices. The market will change to favor one product and disfavor another. This may happen when mass merchandisers enter the market, when the popular press promotes a particular vegetable, or when new medical evidence points to increased or decreased health benefits from certain vegetables.

A good example of changes in a niche market is the salad greens industry. Fifteen years ago, leaf lettuce was almost impossible to find. When leaf lettuces were introduced to the general public, few people accepted them. When chefs in finer restaurants began using them, more affluent people began asking for them in markets. The undersupply led to extremely high prices; as much as $16 per pound was not uncommon. More and more small growers began producing salad greens, but it wasn't until large growers entered the market that the price per pound went down significantly (to $6-10 a pound). Many growers can still get $4-6 a pound for greens, but as more large growers enter the market, this price will continue to drop. Long before the market has bottomed out is when small growers need to diversify and find ways to add value to their crops, like offering pre-cut, washed and ready-to-eat mixed lettuces.

Labor and energy are usually the two greatest greenhouse expenses. If small growers can find a way to decrease costs of either or both of these, their chances of making a profit are strengthened. How can they do this? Cheap sources of energy are key. Sunlight, decomposing compost, and animal heat are three ways to decrease energy costs.

Solar Greenhouses

Greenhouses can be designed to take advantage of solar radiation and cut fuel expenses. Solar greenhouses are popular with small-scale growers. These are super-insulated greenhouses designed to collect and retain solar energy. The technology associated with solar greenhouses is rather detailed. In addition, the literature on solar greenhouses is quite large.

Composting Greenhouses

Heating greenhouses with waste heat generated by compost is a second option that takes advantage of local resources and integrates different farm activities. In a composting greenhouse, heat and carbon dioxide are generated from manure-based compost contained in a special chamber attached to one side of the greenhouse.

Compost-heated greenhouses gained a lot of attention from work undertaken at The New Alchemy Institute at Falmouth, Massachusetts. The New Alchemy Institute was one of the premier appropriate technology centers that operated in the 1970s and 80s. The Institute published widely on ecology, wind energy, solar energy, bioshelters, solar greenhouses, integrated pest management in greenhouses, organic farming, and sustainable agriculture.

Animal-Heated Greenhouses

Small animals like chickens and rabbits produce heat and carbon dioxide in addition to products like eggs and meat. A few growers have taken advantage of this fact and integrate animals with greenhouses as a source of heat. However, it can be a challenge to keep livestock in a greenhouse–the higher temperature and humidity of a greenhouse are generally not healthy for animals.

Anna Edey, mentioned above, uses an "earth-lung" to filter out the toxic ammonia gas from the rabbit and chicken manure she uses in her Solviva greenhouse. In addition, she keeps her chickens in a poultry room attached to the greenhouse where temperatures do not fluctuate from about 70°F.

In 1996, there was a SARE-funded project headed by Rick Meisterheim in Michigan titled "Permaculture Greenhouse System." He raised chickens in the greenhouse and found that they raised the overall temperatures inside the greenhouse by an average of 8°F. This proved to be very helpful in raising vegetable transplants. Rick was able to set out seedlings much earlier than other years, thus increasing the length of time his CSA generated income. Again, bio-heated greenhouses hold interest among small-scale growers who are focused on bio-integrated systems as a form of energy efficiency and biomass utilisation. Of the three greenhouse-heating methods summarised here, solar greenhouses have the least risk and hold the most promise for commercial-scale ventures.

Organic Greenhouse Production

As defined by the USDA in 1980, organic farming is a system that excludes the use of synthetic fertilisers, pesticides, and growth regulators. Organic farmers rely heavily on crop rotations, crop residues, animal manures, legumes, green manures, organic wastes, and mineral-bearing rocks to feed the soil and supply plant nutrients. Insects, weeds, and other pests are managed by mechanical cultivation, and cultural, biological and biorational controls.

Organic certification emerged as a marketing tool during the 1970s and 80s to ensure foods produced organically met specified standards of production. The Organic Foods Production Act, a section of the 1990 Farm Bill, enabled the USDA to develop a national programme of universal standards, certification accreditation, and food labeling. Early in 1998, the USDA released a draft of the new standards for public comment. Public opposition to these proposed standards was vocal, sending a message to the USDA that more work was necessary. A new draft of organic standards may be released in 2000. In the meantime, organic certification will continue to occur at the state level for most of the country.

Growers may choose organic methods for a variety of reasons. One of the attractions of growing organic produce is that it sometimes brings a 10-30% premium in the marketplace. As organically grown produce becomes more commonplace, however, premiums may be the exception rather than the rule, and motivation beyond market premiums should be considered. These may include the possibility of reduced input costs, improved farm safety, reduced environmental impact, and a better functioning agroecosystem.

Greenhouse technology and horticultural practices differ little between conventional and organic greenhouse production. The main variations are concerned with pest control and fertility.

Fertility

Although the process is more complicated, it is possible to obtain adequate nutrients from organic sources, but it takes more careful and creative management. No single fertiliser will provide all of the essential elements required, but a combination of organic products can be devised.

Organic fertilisers have not been well researched in greenhouse vegetable production. However, a 1999 study performed at the University of Kentucky analyzed several products for the levels of nutrients they supplied. The researchers were attempting to prove that organic fertilisers could supply nutrients at the same level as synthetic fertilisers. Products derived from algae (Algamin, a liquid, and Maxicrop, a powder), bat guano, and fish waste (GreenAll Fish Emulsion, a liquid, and Mermaid's Fish Powder) demonstrated nutrient levels comparable to conventional, synthetic fertilisers used for greenhouse plant production.

In 1993, Premier Peat Moss in Canada conducted research on organic wastes from the agri-food industry and their ability to fertilise greenhouse tomato transplants. The researchers found that meal from blood, feathers, meat, crab shells, fish, cottonseed and whey by-products increased shoot weight by 57-83% over non-fertilised plants. The results of their study are shown in the table below.

Table 1: Organic wastes from the agri-food industry

Fertiliser	N- P-K	Shoot wt. (grams)
Crab-shell meal	8.2-1.5-0.5	18.8
Blood meal	12.5-1.1-1.0	18.5
Dried whey sludge	5.3-2.5-0.9	18.3
Feather meal	13.6-0.3-0.2	17.3
Fish meal	10.1-4.5-0.5	17.1
Meat meal	7.7-3.1-0.7	16.3
Cottonseed meal	6.5-1.1-1.6	16.2
Fish-scale meal	10.0-3.7-0.1	15.8
Distiller's dried grains	4.3-0.9-1.1	14.5
Soybean meal	7.5-0.7-2.4	14.4
Wheat bran	2.9-1.4-1.3	13.5
Alfalfa meal	2.5-0.3-1.9	10.8
Canola meal	6.0-1.1-1.3	10.8
None (control)	0-0-0	10.3

Liquid fertiliser applied through irrigation lines–a technique known as fertigation–is a common practice in greenhouse production. Nitrogen and phosphorus are the two primary nutrients injected through fertigation systems. Fertigation can provide supplemental doses of these nutrients, especially at critical periods during the growing season such as fruit filling. However, fertigation is not a substitute for a complete soil-based fertility programme.

The injection of organic fertilisers into low-volume irrigation systems like drip emitter and micro-sprinkler is non-conventional, because the standard method relies on commercial synthetic fertilisers that are highly soluble in water. The major concern with organics is clogging of emitters and pores with particulate matter and algae.

In the early 1990s, researchers in California concluded that fertigation with certain organic fertilisers is feasible. Spray-dried fish and poultry protein fertilisers were injected through various drip, drip-tape, and micro-sprinkler irrigation systems, with even distribution and minimal clogging. The protein products are easy to use because they are finely ground and sold dry. The N-P-K analysis for both fish and poultry protein is 14-0.5-0.7. At the time of the study, the material cost $1.60/lb., or $12/lb. of nitrogen. In comparison, sodium nitrate costs about $2/lb. of nitrogen. For product information on spray-dried protein fertilisers.

Neptune's Harvest®, with fish, seaweed, and fish-seaweed blends, is a product line that is popular with organic growers using drip systems. David Hertzell, an organic greenhouse tomato grower in Kansas, applies Neptune's Harvest through his drip-tape system at 32-oz./400 gallons water every ten days. He and other growers prefer the blended mix to straight seaweed, because they feel the nutrient ratios of trace elements are not as balanced in seaweed alone. David also uses rock phosphate at 10 lbs/85' row, as well as blood meal, bone meal, and alfalfa meal. He uses alfalfa mulch, a heavy application of leaf-mold compost, and cover crops. He also solarises his soil every five years.

James Baker raises organic cucumbers, tomatoes, chives, and squash in Israel. He applies 20-25 cubic meters (706 to 882 cubic feet) of compost per quarter acre in the first year of production, then 10 to 15 cubic meters (353 to 530 cubic feet) on following crops. In addition, he uses feather meal for nitrogen, basalt rock for iron and trace elements, and applies bat guano through his irrigation system. Says Baker, "I'd prefer to use poultry manure, but that's not strictly organic because the chickens are fed antibiotics and hormones, and aren't fed organic grain."

Soil vs. Soilless Culture

Greenhouse production is based on either soil or soilless culture. In soil culture, vegetables are produced in the native soil or in a loamy soil brought inside the greenhouse. Soilless culture, the production of plants in a soilless medium, can be done via hydroponics or with an organic (or artificial) substrate.

The two broad categories of soilless culture are illustrated in the diagram below, with examples of media used in each.

In 1974, 70 percent of U.S. production was based on soil culture and 30 percent on soilless culture. By 1988, a significant shift to soilless systems had occurred, with soil systems making up only 40 percent of the acreage. The majority of large producers in Canada raise tomatoes, cucumbers, and peppers in rockwool, and lettuce is usually raised with a nutrient film technique (NFT) system. Bag culture is often practiced in the U.S., however. The chief reasons soilless culture became so popular are: a) the elimination of fumigation and steam treatments for soil-borne diseases; b) ease of installation; c) technical assistance available for soilless systems; and d) fertility control through commercial soluble fertilisers.

Five greenhouse production methods suitable for organic production are covered in the following material: soil culture, bag culture, vertical towers, straw bale culture and shallow bed culture.

Soil Culture

In soil culture, also known as ground culture, vegetables are raised on level ground as well as in mounded beds. Soil culture is more popular with organic growers than hydroponic methods. One of the reasons organic growers prefer soil culture to hydroponics is that soil-building practices are an already familiar concept based on decades of research and experience.

Production techniques used in ground culture are similar to intensive market gardening methods used in the field. Rear-tine tillers or mechanical spaders are used in soil preparation, and specialised wheel hoes and hand tools are used for cultivation and weed control. Organic fertilisers and soil amendments are extensively used to feed the soil.

In Quebec, organic farmers raising greenhouse tomatoes by the ground culture method use two rows per bed raised about one foot high. Compost is applied at a rate of 1.2 to 2.4 cubic yards per 1000 sq. feet. The first year it is applied 5-6 times per growing season at 5-6 week intervals. In years two through four the compost is applied at a lower rate. Following compost application, the beds are covered with straw mulch. Soil organic matter in these greenhouses ranges from 10-12% up to 25-30%.

In contrast, the Luebkes, one of the best known organic farming families in Austria, fertilise greenhouse tomatoes at a rate of 8 tons/acre/year. However, they use Controlled Microbial Compost™–a high quality, microbially inoculated compost prepared under aerobic (oxygen-present) conditions. In addition to compost, green manures are an integral part of the Luebke soil management programme. After the green manure is mowed but prior to plowdown, the green residue is sprayed with a microbial inoculant to speed decomposition and enhance humus formation. Soils treated in this manner have tested at 15% organic matter.

Cover Cropping in Ground Culture Systems

Cover crops fit nicely into ground culture systems that include a crop-free rest period (in summer, for example). Short-term cover crops include legumes, buckwheat, annual ryegrass, oats, rapeseed, and oilseed radish. Often, the choice of cover crop will depend on the vegetable rotation and the intended purpose. Cover crops can replenish and help build organic matter in soils, increase microbial activity, add nitrogen, and help prevent or suppress diseases and nematodes, especially when brassica and oilseed radish cover crops are used.

Marty Legant, who raises 2 acres of organic winter tomatoes by the ground culture method in Parowan, Utah, has an 8-week period in which cover crops occupy his greenhouse. He's raised buckwheat and oats in the past, and has plans to try hairy vetch. At plowdown, he inoculates the green manure with vitamin B-12, sugar, and a catalyst to help speed decomposition and enhance humus formation.

Anna Edey, whose primary commercial crop is salad greens, recommends the following:

For average soil, add, per square yard: 4 gallons of compost, 4 cubic feet of peat moss, 1-2 pounds of soft rock phosphate, 1-2 pounds of greensand, and lime if needed. Spread it out evenly and blend thoroughly by rototilling twice, rake it smooth and level, and then water well enough to soak through to the full depth.

The next day sow buckwheat seed, spaced ¼ to ½ inch apart, cover lightly and water it well. Let the buckwheat grow for about two to three weeks, to about 10-12 inches high, then turn it under and sow another round of buckwheat. Repeat these buckwheat cycles until the growing beds in the greenhouse are ready to be filled.

In order to maintain highest soil fertility, the soil needs to be enriched with compost and minerals four to five times per year.

An alternative to growing cover crops inside the greenhouse is incorporation of freshly chopped cover crops or pasture-grown forages transported from a nearby field. These are called green-leaf manures.

Weed Control in Ground Culture Systems

Strategies for non-herbicidal weed control inside the greenhouse include a combination of hand hoeing, mechanical cultivation, steam pasteurisation, soil solarisation, mulching (fabric, plastic, or organic mulches), and poultry grazing.

Poultry grazing is a unique method of "biological" weed control that may be an option for some greenhouse growers. This system is used by Lynn Shanks, a vegetable grower near Bixby, Oklahoma. Mr. Shanks raises greenhouse tomatoes and early-season cantaloupes by the ground culture method. To manage weeds, he houses jungle fowl from India in the greenhouse to "graze" on weeds. The jungle fowl are also deft at picking insects off the vegetable plants without damaging any fruits. (Jungle fowl should not be confused with the domesticated chicken, which is notorious for pecking at tomatoes in the garden). The impressive aspect of this system is that weeding requires no labor, mulching, or cultivation on the part of Shanks' family.

Soilless Culture

Soilless culture methods usually have higher yields than ground culture systems.

Another advantage of a soilless system is easier control of disease problems. In the soilless techniques described below, the media can be disposed of after one crop has been grown, or it can be pasteurised relatively easily.

Bag Culture

In bag culture, plants are grown in a soilless medium contained in lay-flat or upright polyethylene bags (2-5 gallons in size) and fed with liquid fertilisers through drip irrigation lines. Media can be peat/vermiculite, sawdust, rockwool, rice hulls, pine bark, peanut hulls, or mixes of any of those. Bag culture is adaptable to both aggregate hydroponics fed with liquid fertilisers as well as standard soilless culture, in which the bulk of fertility is supplied via compost-based potting mixes.

Upright bags should be purchased at greenhouse supply houses, since household garbage bags are usually not strong enough. Upright bags usually contain one plant per bag. Lay-flat bags are closed and are made of 4-mil, UV-protected polyethylene. Lay-flat bags often have two or three plants per bag. Bags are most often placed in double rows in the greenhouse.

Bag culture is well suited to the production of upright and vining crops like tomatoes, cucumbers, and peppers. Bag culture has become the preferred method of greenhouse vegetable production in many parts of the U.S. because it is easy to establish and manage successfully. One problem with bag culture is the potential for excess fertiliser solution to leach out of the drainage holes in the bags and into the greenhouse soil.

Adapting bag culture to standard soilless production in a certified organic greenhouse operation requires specially formulated organic potting mixes. An organic potting mix should not include any synthetic fertilisers or wetting agents. Growers can purchase certified organic potting mixes from a commercial supplier, or mix their own. Organic bag cultured vegetables may include a combination of incorporated, topdressed, liquid-fed, and foliar-fed fertilisation systems.

Adapting bag culture to soluble hydroponics or bioponics requires special hydroponic formulas and ingredients.

Vertical Towers

A form of bag culture is the vertical tower. Long bags full of media are hung from support wires or beams, and plants are placed in slits or holes made on the sides of the bag. The appeal of this system is the very efficient use of greenhouse space; since the plants are using vertical space, very little floor space is needed.

The most popular vertical tower is a patented system called Verti-Gro. Verti-Gro uses square, styrofoam pots, which are stacked 8-10 pots high and can be re-used for 5-10 years. Tim Carpenter, who owns the company, recommends using perlite and coir (coconut fiber) for media. Fertiliser is supplied by a drip irrigation system, in which solution flows from the topmost to the bottom box and into a gravel support base, where it can be captured and re-used. Crops that have been grown with this system include lettuces, strawberries, herbs, tomatoes, greens, spinach, nasturtium, and cut flowers.

Straw Bale Culture

Straw bale culture was an important method of greenhouse cucumber production in Europe and Canada prior to the development of rockwool and NFT hydroponics. Bale culture can be especially useful as a method to avoid soils contaminated with diseases or in greenhouses with concrete or gravel floors. Based on limited experience in Canada in adapting straw bale culture to organics, this method appears to have merit where one of the aforementioned conditions exists.

In bale culture, the greenhouse floor can be concrete or lined with plastic or fabric mulch, and bales are laid out in rows. Bales are pre-wetted with manure tea, which initiates decomposition and causes the bales to heat up. On the average, about 7 gallons of water are required for a normal 50 lb. bale. When the temperature of the bales cools down to below 110° F, they are capped with six inches of a compost-based potting mix to accommodate the root balls of the transplants. Normally, planting would take place on about the eleventh day.

Fertilisers topdressed to the bales include dry bone meal, sul-po-mag, and an organic nitrogen source. Manure tea, seaweed, and fish-seaweed blends are used as liquid fertilisers. As the bales continue to ferment and decompose, heat and carbon dioxide are generated and these byproducts promote plant growth by warming the roots and enriching the greenhouse atmosphere.

In considering straw bale culture, the cost of the bales and topcapping compared with costs of a comparable production system such as bag culture, will need to be weighed. Also, in organic culture, a major factor would be the availability of certified organic straw. Herbicide residues in straw can cause serious injury to greenhouse crops.

Canadian studies conducted in the 1960s and 1970s showed that:

1. Spring-grown cucumbers grown on straw bales yielded an average of 21% more fruit than soil-grown plants, although fall-sown plants did not have a yield increase.
2. Wheat straw bales provided the best results, followed by clover straw. Timothy/alfalfa hay bales gave poor results.
3. Using half a bale per plant caused no significant loss in yield.
4. Using a plastic sheet under the bales caused a 9% reduction in yield.

5. There is little or no difference in yield between bales placed on the soil surface and those buried or half-buried in a trench.
6. Growing in straw bales was economically feasible for spring-sown plants.

Missouri farmer Eric Hemple raises organic arugula, squash, green beans, onions, eggplant, tomatoes, and cucumbers using the straw bale method. A neighbor provides organic straw bales at a price of $3 each. Hemple places the bales between two sheets of corrugated tin braced with stakes to create a more uniform growing surface. He uses a sterile potting mix as his capping medium and bone meal, blood meal, fish emulsion, and rock phosphate as his fertilisers. Hemple uses a drip irrigation system and warns that careful attention to watering is essential, as are proper nutrient levels. He sells all his products at retail and estimates that he is receiving a gross return of 50 cents per square foot of bale space per week.

Shallow Bed Culture

Shallow bed culture, also known as thin layer culture, is the production of sprouts or herbaceous herbs and vegetables in a thin layer–1 to 3 inches in depth–of potting mix or compost laid on top of plastic mulch or woven weed barrier. Herbaceous crops adapted to shallow bed culture include specialty lettuces, greens, cresses, and selected culinary herbs.

At least one commercial sprout grower in California uses this method in the production of organic wheat grass, sunflower, and buckwheat sprouts. Root-zone heating tubes are laid underneath the plastic to provide bottom heat. In turn, they rest on top of polystyrene panels laid directly over the greenhouse soil. Gas-fired water heaters are used to heat the water, but such a system could easily be retrofitted to solar collectors located outside the greenhouse. Narrow boards are laid out every three feet to walk up and down between the beds to seed, water, and harvest.

Jay Fulbright, a commercial lettuce and specialty greens grower in Arkansas, has three growing systems. In the first, he amends his soil with compost and pelletised chicken litter and plants directly into the ground. In the second system, he uses tables made of 2x8 lumber and places 3½" of compost onto 2x4 welded wire mesh, covered with perforated plastic to allow for drainage. All the greens are hand-harvested using this system. The third method involves placing 4" of compost and chicken litter directly onto polypropylene sheets laid on the ground. Jay devised the third method in order to cut down on labor; he hopes to be able to use a salad harvester. He is now harvesting an average of 300 pounds of lettuce per week, which he sells to restaurants and specialty groceries.

Education Concerns for Hunger Organisation (ECHO), a nonprofit organisation that works in agriculture development, is promoting the shallow bed method as a technique for rooftop gardening in urban settings. The beds they describe are deeper-3 to 6 inches in depth-and are more typical of standard trough culture using a soilless mix. They've achieved good success raising a variety of vegetable crops using locally available materials such as leaf mold and compost.

Hydroponics

In hydroponics, plants are grown in a soilless medium and fed with fertiliser nutrients dissolved in solution. There are two broad categories of hydroponics: liquid hydroponics and aggregate hydroponics.

Liquid systems feature the nutrient film technique (NFT), aeroponics, floating raft, and noncirculating water culture. Aggregate systems include inert, organic, and mixed media contained in bag, trough, trench, and bench setups.

Liquid systems like NFT are often managed as closed, or recirculating systems. Aggregate systems are commonly managed as open, or flood and drain fertigation systems.

There are three basic approaches to organic hydroponic vegetable production: a) soluble organics, also known as hydro-organics, b) integration of hydroponics with aquaculture, also known as aquaponics, and c) bioponics.

Hydro-organics is based on hydroponic solutions derived from organic fertilisers like fish meal, spray-dried blood, and guano. Aquaponics is based on recirculating aquaculture in which irrigation effluent from fish tanks is used to fertigate sand or gravel cultured hydroponic vegetables.

The key difference between organic and chemical hydroponics is the source of fertility and presence of microorganisms. Microbes are essential to organic systems because they help regulate pH and availability of nutrients. In addition, bioponic researchers claim that enzymes produced by the microbes interact with plant roots and stimulate biochemical processes in the plants which enhance the flavor of bioponic produce.

Variety Selection

Variety or cultivar selection plays an important role in greenhouse vegetable production. Cultivars often originate in greenhouse breeding programmes where selection is based on disease resistance and ability to perform under greenhouse climatic conditions. Greenhouse seed vendors and the Cooperative Extension Service can recommend appropriate cultivars.

Economics of Greenhouse Vegetable Production

Greenhouses are expensive to build and operate. A commercial greenhouse (30' x 100') with complete heating, cooling, and ventilation systems will cost between $10,000 to $30,000 to erect and equip.

Low cost greenhouses–like hoop houses and attached solar greenhouses–can be constructed for as little as $500 to $1,500. Hoop houses are similar to plastic Quonset greenhouses but are not equipped to heat or cool, and are used primarily for season extension. Solar greenhouses work fine for small-scale production–especially for the production of transplants and cool-season lettuces–but may not provide adequate space or climate control for commercial tomato, cucumber, or pepper production.

Otto Wells, a professor at the University of New Hampshire, found that growing tomatoes in an unheated 14x96-foot hoop house in New Hampshire could be profitable, if he realised a yield of 2000 pounds and sold the tomatoes for $1.60/lb. Using these estimates, the grower would earn a net return of 71 cents per pound.

The greenhouse vegetable business is competitive. Wholesale markets for off-season produce often become saturated and cause produce prices to drop. Competition stems from an established domestic industry as well as Dutch imports. In addition, field-grown Mexican produce frequently sells at half the price of greenhouse produce. Local and specialty markets provide alternative outlets, but the volume of produce that can be moved is much less than what can be shipped to distant markets.

Greenhouse vegetable yields determine potential gross sales. Typical yields of greenhouse tomatoes are 20 to 30 lbs. per vine, or 2-3 lbs. per square foot. Greenhouse cucumbers yield around 2 dozen fruits per vine. Greenhouse peppers yield 2½ -3 lbs./ sq. ft. A study conducted in Missouri in the winter of 1995-96 showed that supplemental lighting of tomatoes increased total yields from 12,444 kg to 18,840 kg. Because the lighted tomatoes were larger, they brought a better price and resulted in additional revenues of $25,000.

Prior to sinking lots of money into a greenhouse venture, growers should examine produce prices in their region and estimate their cost of production. Historically, the breakeven price for most greenhouse tomatoes has been around 75 cents per pound, with selling prices ranging from 90 cents to $1.60 per pound. The break-even price for cucumbers is similar–around 75 cents per pound.Eliot Coleman and his wife Barbara Damrosch raise organic produce in their greenhouses in Maine. They charge $6.25/lb for salad mix, $3.00/lb. for bulk spinach, $1.25-$2.00 for carrots, and $3.00/lb for baby turnips. They dropped scallions because they couldn't make money from them.Estimates of net income from conventional greenhouse tomatoes range from $3,100 to $18,500 per greenhouse unit. These estimates are for good yields and favorable market conditions. Low yields, or a dip in the market, can lead to negative returns to the grower.

Developments in the Greenhouse Industry

One exciting development is the introduction of bumblebees for greenhouse tomato pollination. Pollinating by hand is expensive (in terms of labor), tedious, and time-consuming. Several companies are now raising bumblebees to be used for indoor pollination. Although the hives are expensive, they pay for themselves many times over. There have also been many new biorational pesticides in the last five years. Some of these biorational pesticides (also known as least-toxic, environmentally friendly, or biopesticides) include *Bt*, *Beauveria bassiana*, and Cinnamite™, which was just released in 1999 for control of several greenhouse pests.

Use of biological control agents has also increased in the last decade. Much more is known about how to use these insects to their fullest advantage and about how to apply insecticides while preserving predators. Additionally, new cultivars bred exclusively for greenhouse growing are now available to small growers. Seed vendors like DeRuiter's are constantly testing new cultivars for greenhouse production. Although seed for greenhouse crops is still more expensive than for field crops, it has become more affordable. Growers themselves have been testing cultivars for suitability in greenhouse work, especially in the area of heirloom vegetables.

Another change that is occurring is in the field of greenhouse coverings. Modern greenhouse plastics are resistant to breakdown from ultraviolet light and are rated to last several years. This reduces annual expenses for materials, labor to reglaze, and fees for landfill disposal. Structured sheets are the deluxe alternative to Quonset-style plastic houses. They are more expensive to purchase and install, but they provide optimum environmental controls and have a very long life. Research into the chemistry associated with greenhouse glazing is aimed at energy savings, pest control, and enhancement of plant growth.

REFERENCES

Al-Abdulkador, Ahmed M. 1992. *Economic Feasibility of Greenhouse Vegetable Production in Oklahoma*. Oklahoma State University, Stillwater. M.S. Thesis. 139 p.

Anderson, Robert G. and Robert Hadad. 1999. "Nutrient analysis of organic fertilisers for greenhouse vegetable production". *HortScience*. June. p. 463. Abstract.

Byczynski, Lynn. 1993. "Greenhouse tomato considerations". *Growing for Market*. April. p. 1, 12.

Census of Agriculture. 1997. "U.S. Census of Horticulture Specialties". Bureau of the Census, U.S. Department of Commerce, Washington, D.C.

Edey, Anna. 1998. "Solviva: How to Grow $500,000 on One Acre and Peace on Earth". Trailblazer Press, Martha's Vineyard, MA. 221 p.

Flaim, John. 1999. "Growing in haybales. *Small Farm Today*. August. p. 58-60.

Gagnon, Bernard and Sylvain Berrouard. 1994. "Effects of several organic fertilisers on growth of greenhouse tomato transplants". Canadian Journal of Plant Science. January. p. 167-168.

Hochmuth, George J. 1992. "Production and economics of rockwool tomatoes in Florida". p. 40-46. In: Proceedings of the 13th Annual Conference on Hydroponics. Hydroponic Society of America, Concord, CA.

Jones, J. Benton, Jr. 1999. *Tomato Plant Culture*. CRC Press, Boca Raton, FL. p. 90.

5

Molecular Plant Breeding

Plant breeding describes methods for the creation, selection, and fixation of superior plant phenotypes in the development of improved cultivars suited to needs of farmers and consumers. Primary goals of plant breeding with agricultural and horticultural crops have typically aimed at improved yields, nutritional qualities, and other traits of commercial value. The plant breeding paradigm has been enormously successful on a global scale, with such examples as the development of hybrid maize, the introduction of wheat (Triticum aestivum) and rice (Oryza sativa) varieties that spawned the Green Revolution, and the recent commercialisation of transgenic crops. These and many other products of plant breeding have contributed to the numerous benefits global society has received from greater sustainable supplies of carbon that may be harvested as food, feed, forests, fiber, and fuel.

Plant breeding has a long history of integrating the latest innovations in biology and genetics to enhance crop improvement. Prehistoric selection for visible phenotypes that facilitated harvest and increased productivity led to the domestication of the first crop varieties and can be considered the earliest examples of biotechnology. Darwin outlined the scientific principles of hybridisation and selection, and Mendel defined the fundamental association between genotype and phenotype, discoveries that enabled a scientific approach to plant breeding at the beginning of the 20th century. Despite the immediate recognition among some plant breeders of the importance of Mendelian genetics, full integration was delayed for nearly 20 years until quantitative genetics reconciled Mendelian principles with the continuous variation observed for most traits considered important by most plant breeders. Subsequent advances in our understanding of plant biology, the analysis and induction of genetic variation, cytogenetics, quantitative genetics, molecular biology, biotechnology, and, most recently, genomics have been successively applied to further increase the scientific base and its application to the plant breeding process.

The plant biotechnology era began in the early 1980s with the landmark reports of producing transgenic plants using Agrobacterium. Molecular marker systems for crop plants were developed soon thereafter to create high-resolution genetic maps and exploit genetic linkage between markers and important crop traits. By 1996, the commercialisation of transgenic crops demonstrated the successful integration of biotechnology into plant breeding and crop improvement programmes. As depicted in Figure 1, introgression of one or a few genes into a current elite cultivar via backcrossing is a common plant breeding practice. Methods for markerassisted backcrossing were developed rapidly for the introgression of transgenic traits and reduction of linkage drag, where molecular markers were used in genome scans to select those individuals that contained both the transgene and the greatest proportion of favorable alleles from the recurrent parent genome. During the past 25 years, the continued development and application of plant biotechnology, molecular markers, and genomics has established new tools for the creation, analysis, and manipulation of genetic variation and the development of improved cultivars.

PRINCIPLES AND PRACTICES OF MOLECULAR PLANT BREEDING

Breeding Schemes and the Genetic Gain Concept

Conceptually, plant breeding is simple: cross the best parents, and identify and recover progeny that outperform the parents. In practice, plant breeding is a three step process, wherein populations or germplasm collections with useful genetic variation are created or assembled, individuals with superior phenotypes are identified, and improved cultivars are developed from selected individuals. A wide diversity of approaches, tailored to the crop species and breeding objectives, have been developed for improving cultivars. These breeding methods feature different types of populations, selection procedures, and outcomes.

Figure 1 summarises the three breeding methods that are commonly employed in crop improvement programmes. As mentioned previously,when the goal is to upgrade an established elite genotype with trait(s) controlled by one or a few loci, backcrossing is used either to introgress a single gene (Fig. 1A) or to pyramid a few genes (Fig. 1B). For genetically complex traits, germplasm improvement instead requires reshuffling of the genome to produce new favorable gene combinations in the progeny. The pedigree breeding method produces such novelty via crossing and recombination among superior, yet complementary, parents and selection among segregating progeny for improved performance (Fig. 1C). Recurrent selection aims to simultaneously increase the frequencies of favorable alleles at multiple loci in breeding populations through intermating of selected individuals (Fig. 1D). For hybridised crops such as maize, recurrent selection may be extended to improve the performance of distinct complementary populations (e.g. heterotic groups) that are used as parents to form superior hybrid combinations. This practice is referred to as reciprocal recurrent selection.

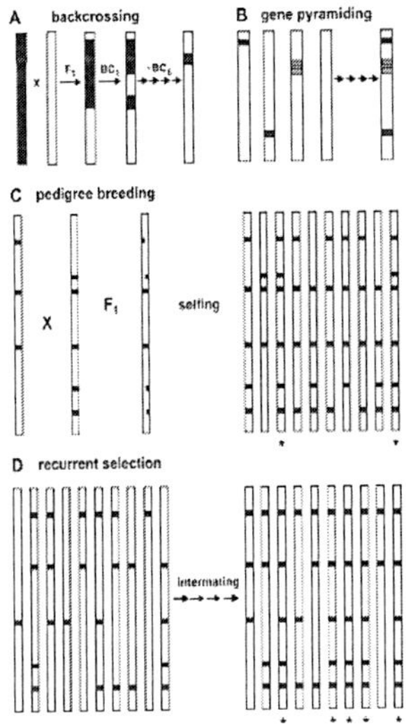

Figure 1. Common breeding and selection schemes.

Quantitative genetic principles have been particularly powerful as the theoretical basis for both population improvement and methods of selecting and stabilising desirable genotypes. An important concept in quantitative genetics and plant breeding is genetic gain (DG), which is the predicted change in the mean value of a trait within a population that occurs with selection. Regardless of species, the trait of interest, or the breeding methods employed, DG serves as a simple universal expression for expected genetic improvement.

Figure 2 shows the genetic gain equation and an expansion of its terms to fundamental parameters of quantitative genetics.

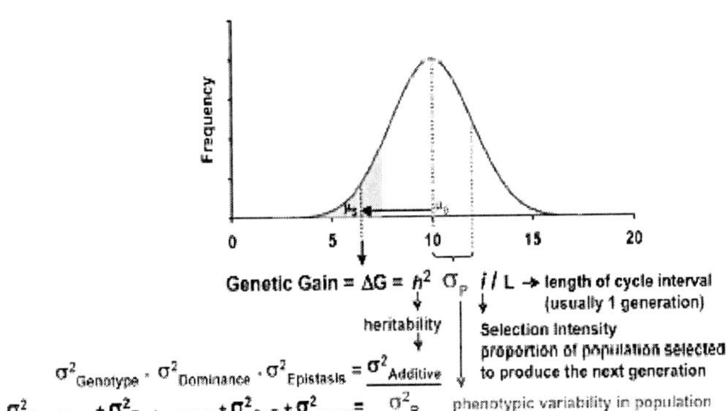

Figure 2. The genetic gain equation and its component variables.

Though clearly an oversimplifi-cation of the advanced quantitative genetic principles employed in plant breeding, the genetic gain equation effectively relates the four core factors that influence breeding progress: the degree of phenotypic variation present in the population (represented by its SD, s_p), the probability that a trait phenotype will be transmitted from parent to offspring (heritability, h^2), the proportion of the population selected as parents for the next generation (selection intensity, i, expressed in units of SD from the mean), and the length of time necessary to complete a cycle of selection (L). L is not only a function of how many generations are required to complete a selection cycle, but also how quickly the generations can be completed and how many generations can be completed per year.

It is clear that ΔG can be enhanced by increasing σ_p, h^2, or i, and by decreasing L. Thus, the genetic gain equation provides a framework for comparing the predicted effectiveness of particular breeding strategies and is often used as a guide to the judicious allocation of resources for achieving breeding objectives. When considered in the context of the genetic gain concept, molecular plant breeding offers powerful new approaches to overcome previous limitations in maximising ΔG.

Genetic Diversity for Crop Improvement

The maximum potential for genetic gain is proportional to the phenotypic variation (σ_p) present in the original source population and maintained in subsequent cycles of selection. Phenotypic variation is positively associated with genetic diversity, yet also depends on environmental factors and the interactions between genotype and environment.

Genetic diversity may be derived from breeding populations (either naturally occurring or synthetic), segregating progeny from a cross of selected parental lines, exotic materials that are not adapted to the target environment, wide interspecific crosses, naturally occurring or induced mutations, the introduction of transgenic events, or combinations of these sources. However, not all phenotypic variation is equal. For example, the use of exotic germplasm has been extremely successful for improving many crop species, but difficulties may be encountered through the introduction of undesirable alleles associated with lack of adaptation. The need for genetic diversity must be balanced by elite performance, because choosing the best parents is key to maximising the probability for successful improvement.

In contrast, the expected increase in linkage disequilibrium among elite populations derived from intense prior selection may also limit the creation of new genetic combinations for future gain. Intermating source populations for genetic recombination may overcome this problem, but delays cultivar development. Molecular markers and more recently, high-throughput genome sequencing efforts, have dramatically increased knowledge of and ability to characterise genetic diversity in the germplasm pool for essentially any crop species. Using maize as one example, surveys of molecular marker alleles and nucleotide sequence variation have provided basic information about genetic diversity before and after domestication from its wild ancestor teosinte, among geographically distributed landraces, and within historically elite germplasm.

This information enriches investigations of plant evolution and comparative genomics, contributes to our understanding of population structure, provides empiricalmeasures of genetic responses to selection, and also serves to identify and maintain reservoirs of genetic variability for future mining of beneficial alleles. In addition, knowledge of genetic relationships among germplasm sources may guide choice of parents for production of hybrids or improved populations. While molecular markers and other genomic applications have been highly successful in characterising existing genetic variation within species, plant biotechnology generates new genetic diversity that often extends beyond species boundaries.

Biotechnology enables access to genes heretofore not available through crossing and creates an essentially infinite pool of novel genetic variation. Genes may be acquired from existing genomes spanning all kingdoms of life, or designed and assembled de novo in the laboratory. Both subtle and extreme examples of the power of transgenes to introduce novel phenotypic variation can be found in the three different transgenes developed for resistance to glyphosate herbicides in maize and other crops.

The first glyphosate-tolerant maize hybrids used a modi-fied version of the endogenous maize gene encoding 5-enol-pyruvylshikimate-3-P synthase, which was followed later by events produced with a 5-enol-pyruvylshikimate-3-P synthase gene isolated from Agrobacterium. More recently, a synthetic gene with enhanced glyphosate acetyltransferase activity was created via gene shuf-fling and selection in a microbial system. Each of these glyphosate-tolerant maize events also illustrates another benefit of biotechnology, where new combinations of regulatory sequences (e.g. the cauliflower mosaic virus 35S and rice actin1 promoters) may be used to achieve optimal trait expression with respect to overall activity and tissue distribution relative to what might be possible with endogenous genes.

Molecular Plant Breeding Increases Favorable Gene Action

Quantitative genetics uses the theoretical concept of heritability to quantify the proportion of phenotypic variation that is controlled by genotype. In practice, heritability is greatly influenced by the genetic architecture of the trait of interest, which is described by the number of genes, the magnitude of their effects, and the type of gene action associated with phenotypes. Better knowledge of genetic architecture and favorable gene action (that which is more amenable to selection) often has the greatest impact on improving genetic gain. For the genetic gain formula, heritability (h^2) is used in its narrow sense, representing the proportion of phenotypic variation due to additive genetic effects (those that reflect changes in allele dosage or allelic substitutions). Additive genetic effects are also referred to as the breeding value because they are predictably transmitted to progeny. Deviations from additive effects are significant for many traits, and are partitioned into either dominance effects that reflect the interactions between different alleles at the same locus or epistatic effects resulting from interactions among different loci. Gene action and breeding values are characterised by progeny testing, where the phenotypes of individuals in a population are compared to their parents and siblings produced from either self-pollination or outcrossing.

Previous efforts to develop large numbers of molecular markers, high density genetic maps, and appropriately structured mapping populations have now made routine for many crop species the ability to simultaneously define gene action and breeding value at hundreds and often thousands of loci distributed relatively uniformly across entire genomes. The results from such mapping studies provide greatly improved estimates for the number of loci, allelic effects, and gene action controlling traits of interest. More importantly, genomic segments can be readily identified that show statistically significant associations with quantitative traits (quantitative trait loci [QTLs]). In addition to genetic mapping in families derived from biparental crosses, new advances in association genetics with candidate genes and approaches that combine linkage disequilibrium analysis in families and populations further enhance power for QTL discovery.

Information about QTLs can be used in a number of ways to increase heritability and favorable gene action. For traits exhibiting low to moderate heritability, such as grain yield, QTLs, and their associated molecular markers often account for a greater proportion of the additive genetic effects than the phenotype alone. Furthermore, knowledge of genetic architecture can be exploited to add or delete specific alleles that contribute to breeding value. When either genetic linkage or epistasis among loci with antagonistic effects on a trait limits genetic gain, QTL information can be used to break these undesirable allelic relationships. Success in using information about QTLs to increase genetic gain depends greatly on the magnitude of QTL effects, precise estimation of QTL positions, stability of QTL effects across multiple environments, and whether QTLs are robust across relevant breeding germplasm. Prediction of QTL positions is enhanced by further fine mapping, which facilitates testing QTL effects and breeding values in additional populations. When the density of observed recombinations approaches the resolution of single genes, the causal genetic change for aQTLcan be determined. Molecular isolation of QTLs permits the development of perfect or functional molecular markers at the potential resolution of the fundamental unit of inheritance, the nucleotide, and dramatically increases the specificity and precision by which genetic effects are estimated and manipulated in breeding programmes.

The use of transgenes can further simplify the genetic architecture for desirable traits, in ways that may be superior to or not possible even when perfect markers are available for robust QTLs of large effect. Transgenes typically condition strong genetic effects at operationally single loci,which also exhibitdominant gene action where only one copy of the event is needed for maximal trait expression in a hybrid cultivar. These features of transgenes can reduce complex quantitative improvement to a straightforward, often dramatic, solution. Excellent examples are provided by the expression in transgenic corn hybrids of insecticidal toxin proteins from Bacillus thuringiensis (Bt) to reduce feeding damage by larvae of the European corn borer (Ostrinia nubilalis) or the corn rootworm beetle (Diabrotica spp.). Partial resistance in maize germplasm to these insect pests had been previously characterised as quantitatively inherited traits with low heritability, but the Bt transgenic events offer a simply inherited alternative that is efficiently manipulated in breeding programmes.

By simplifying genetic architecture, transgenes may also permit disruption of allelic interactions between factors controlling the trait of interest and other important performance characteristics. For example, employing a transgenic source of insect resistance (e.g. a single locus Bt transgene) may facilitate selection for favorable alleles for yield improvement that are tightly linked in repulsion with endogenous genes for resistance to the same class of insect pests. In addition, transgenic events may be engineered to uncouple negative pleiotropic effects from beneficial phenotypes conditioned by recessive mutations. This application is illustrated by the use of RNA interference to specifi-cally down-regulate zein seed storage protein gene expression. This strategy mimics the effects of the *opaque2* mutation on improving the amino acid profile of maize grain for animal feed, while circumventing the softer endosperm texture and susceptibility to fungal pathogens typically associated with *opaque2*. Transgenic events can also be designed to intervene at key regulatory steps for entire metabolic or developmental pathways, such that gene action for the corresponding traits are largely inherited as single dominant factors that are less sensitive to environmental effects. Examples include the expression of a transcription factor that increases drought tolerance, and altering the balance between levels of the GLOSSY15 transcription factor relative to its repressor, *microRNA172*, to delay flowering time in maize hybrids.

Biotechnology also facilitates the molecular stacking of transgenes that control a trait or suite of traits into a single locus haplotype defined by a transgenic event. Examples of such an approach include the initial Golden Rice, recently released Yield-Guard VT Triple transgenic maize hybrids where herbicide tolerance and multiple insect resistance traits are integrated as one genomic locus, or the combination of transgenes that simultaneously increase synthesis and decrease catabolism of Lys in maize seeds.

Molecular Plant Breeding Increases the Efficiency of Selection

Conventional plant breeding that relies only on phenotypic selection has been historically effective. However, for some traits, phenotypic selection has made little progress due to challenges in measuring phenotypes or identifying individuals with the highest breeding value. The effects of environment, genotype by environment interaction, and measurement errors also contribute to observed differences. Evaluation of genotypes in multiple environments with replicated designs allows better estimation of breeding values, but requires additional time and expense. For some traits, it may be necessary to sacrifice the individual to measure phenotypes, or trait expression may depend on variable environmental conditions (e.g. disease pressure) and the stage of development (e.g. grain quality can only be assessed after flowering). Furthermore, plant breeders typically must simultaneously improve a suite of commercially valuable traits, which may limit gains from selection. Just as molecular plant breeding helps to expand genetic diversity, characterise genetic architecture, and modify gene action, its methods can also be applied to increasing the effi-ciency of selection.

An extensive body of literature has considered the utility of molecular marker-assisted selection and its fit with different breeding methods (Fig. 1). Molecular marker genotypes that are either within genes or tightly linked to QTL influencing traits under selection can be employed as a supplement to phenotypic observations in a selection index. In cases where genetic correlations are high, further effi-ciencies can be gained by substituting genotypic for phenotypic selection during some selection cycles, which can reduce phenotyping efforts and cycle times by permitting the use of off-season nurseries. Johnson summarised an early example of combining phenotypic data and molecular marker scores to increase selection gains for maise grain yield and resistance to European corn borer. An effective strategy to simultaneously modify multiple traits is the use of selection indices that consider multiple factors in choosing the final improved genotype. Eathington et al. recently reported on results obtained from the use of multiple trait indices and markerassisted selection for nearly 250 unique corn breeding populations. Use of molecular markers increased breeding efficiency approximately 2-fold relative to phenotypic selection alone, with similar gains also observed in soybean (Glycine max) and sunflower (Helianthus annuus) populations.

Marker-assisted selection can also significantly enhance genetic gain for traits where the phenotype is difficult to evaluate because of its expense or its dependence on specific environmental conditions. Molecular markers may be used to increase the probability of identifying truly superior genotypes, by focusing testing resources on genotypes with the greatest potential (i.e. early elimination of inferior genotypes), by decreasing the number of progeny that must be screened to recover a given level of gain, and by enabling simultaneous improvement for traits that are negatively correlated. Successful examples include resistance to soybean cyst nematode, resistance to cereal diseases, and drought tolerance in maise. The efficiency of phenotypic selection for some complex traits can be enhanced by including physiological or biochemical phenotypes as secondary traits, if these exhibit strong genetic correlations with the target trait and possess high heritability. Recent advances in functional genomics permit the populationscale profiling of RNA abundance, protein levels and activities, and metabolites that are associated with important traits. In addition to molecular markers that tag DNA sequence variation, such genetical genomics approaches may provide additional secondary phenotypes as selection targets, particularly for traits defined by responses to environmental, developmental, or physiological cues. Marker-assisted selection also accelerates the deployment of transgenes in commercial cultivars. Typically, this has been achieved through marker-assisted backcrossing. However, for future biotechnology improvements such as tolerance to drought or nutrient limitation, forward breeding may be required to cooptimise transgene expression and genetic background because endogenous genes and environmental factors may have the potential to influence the phenotypes resulting from transgenic modifications. Of course, use of molecular markers could aid forward breeding efforts as well. Alternatively, discovery efforts for additional genes or QTLs that are necessary for dependable trait performance may suggest design of new transgene constructs that stack primary transgenes with known genetic modifiers into secondgeneration transgenic events.

INCREASING ADOPTION OF MOLECULAR PLANT BREEDING

The adoption of molecular plant breeding approaches has occurred at different rates among crop species and institutions engaged in crop improvement, due to the combined influence of scientific, economic, and sociological factors. Important early scientific barriers included the recalcitrance of cereal crop species to *Agrobacterium*-mediated transformation and lack of knowledge about genetic control of traits already defined as important breeding targets. Continued research and technology development has largely overcome obstacles for plant transformation of nearly all important crop and horticultural species. Similarly, information gained from genomics research in plant species and other organisms has generated a wealth of information about gene structure and function, as well as large numbers of molecular markers for use in plant breeding. Despite these resources, genetic specificity of robust QTLs remains elusive, unless breeding programmes and associated information management systems are restructured to fully integrate knowledge of pedigrees, phenotypes, and marker genotypes that can be leveraged to optimise response to selection. Even with such integration, modifying regulatory functions remains a scientific challenge for molecular breeding because it is difficult to determine the sequence basis for regulatory changes and to predict their phenotypic effects.

Once enabling technologies in biotechnology and genomics become available, economic factors often dictate the degree to which these innovations are integrated into existing plant breeding programmes. The expense of gaining governmental regulatory approval for commercial release of transgenic varieties is a significant economic barrier. The costs associated with the development, establishment, and operation of molecular plant breeding are greater than conventional plant breeding practices, requiring significant investments in new research infrastructure and intellectual capacity. Such resources initially existed only in private agribusiness firms and a handful of larger public institutions, further accelerating an ongoing trend for increased industrialisation of plant breeding programmes among major crops such as corn, soybeans, cotton (Gossypium hirsutum), and wheat. Where there has been adoption in companies, the balance may favor biotechnology over QTLs for improving complex traits, despite greater product development costs, because transgenics can be designed to produce stronger, even dramatic, phenotypic effects, and can often be more rapidly deployed across a broader range of germplasm, resulting in new solutions sooner.

Though molecular breeding is now considered an essential component of current crop improvement efforts for major crops by large companies, the broad applicability of modern molecular approaches to conventional plant breeding remains a source of debate among some practicing plant breeders in the public sector, particularly for minor crops. In addition to the valid scientific and economic factors that have delayed or prevented adoption of molecular approaches in meeting some plant breeding objectives, there are at least three additional reasons contributing to this view. First, molecular plant breeding requires training and expertise in both molecular biology and plant breeding. Educational efforts that delivered such interdisciplinary training were initially established in the early 1990s, but still remain limited to a relatively small group of academic institutions with historic strengths in plant breeding. A second reason for reduced enthusiasm to embrace biotechnology among some plant breeders is the problems with acceptance of transgenic crops among certain governments and groups of consumers, as exemplified by the shelving of wheat varieties with transgenes for resistance to glyphosate herbicides. Finally, excitement about the potential of molecular plant breeding also stimulated shifts in funding at public institutions to enhance intellectual capacity and infrastructure for molecular genetics and genomics research, which ironically often occurred at the expense of conventional plant breeding. This emphasis may have been temporarily necessary to establish the foundations for 21st century plant biology, but there is currently a growing recognition that increased investment in plant breeding capacity and translational research linking molecular methods with breeding objectives is necessary to fully realise the potential of recent advances in biotechnology and genomics.

REFERENCES

Baenziger PS, Russell WK, Graef GL, Campbell BT (2006) "Improving lives: 50 years of crop breeding, genetics, and cytology" (C-1). *Crop Sci* 46: 2230–2244

Duvick DN (2001) "Biotechnology in the 1930s: the development of hybrid maize" . *Nat Rev Genet* 2: 69–74

Guner N, Wehner TC (2003) "Survey of U.S. land-grant universities for training of plant breeding students" . *Crop Sci* 43: 1938–1944

Hallauer AR (2007) "History, contribution, and future of quantitative genetics in plant breeding: lessons from maize" . *Crop Sci* 47: S4–S19

Jansen RC, Nap JP (2001) "Genetical genomics: the added value from segregation" . *Trends Genet* 17: 388–391

6

Marker-assisted Selection in Crop breeding

Conventional plant breeding is primarily based on phenotypic selection of superior individuals among segregating progenies resulting from hybridization. Although significant strides have been made in crop improvement through phenotypic selections for agronomically important traits, considerable difficulties are often encountered during this process, primarily due to genotype—environment interactions. Besides, testing procedures may be many times difficult, unreliable or expensive due to the nature of the target traits (e.g. abiotic stresses) or the target environment.

Molecular marker-assisted selection, often simply referred to as marker-assisted selection (MAS) involves selection of plants carrying genomic regions that are involved in the expression of traits of interest through molecular markers. With the development and availability of an array of molecular markers and dense molecular genetic maps in crop plants, MAS has become possible for traits both governed by major genes as well as quantitative trait loci (QTLs).If the individual gene(s) or QTLs with significant influence on specific target trait(s) can be identified based on their linkage to molecular makers, the efficiency of incorporating the desired traits in elite germplasm could be greatly enhanced. While plenty of information is already available on different kinds of marker systems and gene–marker associations, the practicalities of designing an MAS strategy, increasing their chances of success and efficiency have not received adequate attention till now. A number of molecular investigations have revealed useful gene–marker associations in different crops, but routine implementation in ongoing plant breeding programmes is still in its infancy. It is an opportune time to discuss the practical aspects of designing an MAS strategy and high throughput genotyping techniques that increase their efficiency significantly.

SALIENT REQUIREMENTS FOR MAS

In general, the success of a marker-based breeding system depends on three main factors: (i) A genetic map with an adequate number of uniformly-spaced poly-morphic markers to accurately locate desired QTLs or major gene(s); (ii) Close linkage between the QTL or a major gene of interest and adjacent markers; (iii) Adequate recombination between the markers and rest of the genome; and (iv) An ability to analyse a larger number of plants in a time- and cost- effective manner.

The success of MAS depends on location of the markers with respect to genes of interest. Three kinds of relationships between the markers and respective genes could be distinguished; (i) The molecular marker is located within the gene of interest, which is the most favourable situation for MAS and in this case, it could be ideally referred to as gene-assisted selection. While this kind of relationship is the most preferred one, it is also difficult to find this kind of markers. For instance, microsatellite or simple sequence repeat (SSR) markers have been designed using the available DNA sequence information for the *opaque2* allele that confers high lysine and tryptophan content in the maize kernel. This has offered an efficient means of tracking the *opaque2* allele in breeding for nutritionally superior maize genotypes, since the marker is located within the gene sequence itself and cosegregates with the target gene. (ii) The marker is in linkage disequilibrium (LD) with the gene of interest throughout the population. LD is the tendency of certain combination of alleles to be inherited together. Population-wide LD can be found when markers and genes of interest are physically close to each other. Selection using these markers can be called as LD-MAS. (iii) The marker is in linkage equilibrium (LE) with the gene of interest throughout the population, which is the most difficult and challenging situation for applying MAS.

However, in most cases, especially for the polygenic traits, the target gene(s) within a QTL have not been characterised at the molecular level. Therefore, genomic regions to be selected using MAS are often chromosome segments carrying QTLs in case of polygenic traits. It is preferable either to have two polymorphic DNA markers flanking the target gene (or a QTL), or a marker within a QTL (if the chromosome segment is more than 20 cM) to eliminate the possibility of genotypes presenting a double recombination between the two flanking markers. Depending on the nature of the genomic region (cloned gene, major or minor QTL) involved in the expression of a target trait and the number of selected QTLs or genomic regions that need to be manipulated, several MAS schemes have been proposed that will be discussed later.

In the context of MAS, DNA-based markers can be effectively utilised for two basic purposes: (i) tracing favourable allele(s) (dominant or recessive) across generations and (ii) identifying the most suitable individual(s) among the segregating progeny, based on allelic composition across a part or the entire genome.

FOREGROUND AND BACKGROUND SELECTION

Molecular markers are now increasingly being employed to trace the presence of target genes (foreground selection) as well as for accelerating the recovery of the recurrent parent genome (background selection) in backcross pro-grammes. Marker-assisted backcrossing (MAB) improves the efficiency of backcross breeding in three ways: (i) If the phenotype of the desired gene cannot be easily assayed, backcross (BC) progeny possessing a marker allele from the donor parent at a locus near/within the target gene can be selected with a good probability of carrying the gene; (ii) Markers can be used to select BC progeny with least amounts of donor parent germplasm in the genome outside the target region; and (iii) Markers can be used to select rare progeny that are the result of recombination near the target gene, thus minimising the effects of linkage drag.

Transfer of recessive genes through conventional breeding requires additional selfing generations after every backcross, a procedure that is prohibitively slow for most commercial breeding purposes. Marker-assisted foreground selection was effectively utilised for introgression of disease resistance genes by Melchinger, who presented an a priori approach for calculating the minimum number of individuals and family size required in recurrent backcrossing. However, due to lack of allele-specific markers, practical examples of this approach in plant breeding are limited. One successful example is the conversion of normal maize lines into quality protein maize (QPM) through marker-assisted transfer of a recessive mutant allele, *opaque2*, using allele-specific molecular markers. In animal breeding, availability of an array of allele-specific markers has been facilitating applications of this approach on a commercial scale to eliminate disease and stress-susceptibility genes.

'Marker-assisted background selection', a term coined by Hospital and Charcosset, was initially proposed by Young and Tanksley, and experimented by various researchers. This strategy has been used extensively in commercial maize breeding programmes, particularly for selection of lines carrying transgenes conferring herbicide tolerance or insect resistance. Several parameters need to be optimised in the background selection programmes. Flanking markers for the target allele are necessary to remove linkage drag. The optimal distance between the target gene and flanking markers governs the selection intensity that can be exerted. The equations given by Hospital *et al.* and Frisch *et al.* are helpful in determining the number of BC plants that need to be generated and typed with a special set of flanking markers.

MAS FOR IMPROVEMENT OF QUALITATIVE TRAITS

When the expression of a target trait is controlled by a single gene or by a gene that accounts for a high proportion of the phenotypic variance of the trait, transfer of that specific gene from the donor to the recipient line can lead to significant improvement of the trait. For the introgression of qualitative traits such as pathotype-specific disease resistance or certain quality characteristics, which are typically governed by single dominant/recessive genes, backcross breeding has been used for a long time.

Traditional backcrossing programmes are planned on the assumption that the proportion of the recurrent parent genome is recovered at a rate of $1-(1/2)t+1$ for each of t generations of backcrossing. Thus, after four backcrosses, we expect to recover $1-(1/2) = 96.9\%$ of the recurrent parent genome. However, any specific BC progeny will deviate from this expectation due to chance and to linkage between the gene from the donor parent being selected for and nearby genes. A good example of surprising amount of the linkage drag that accompanies backcross breeding programmes was reported by Young and Tanksley, who genotyped the chromosome carrying the *Tm2* disease resistance gene in several tomato cultivars that were developed by introgressing the gene from a wild relative, *Lycopersicon peruvianum* via backcross breeding. They found that even cultivars developed after 20 backcrosses contained introgressed segments as large as 4 cM and one cultivar developed after 11 backcrosses still contained the entire chromosome arm carrying the gene from the donor parent. Besides, for the transfer of a single dominant gene, a minimum of six backcross generations would normally be required to recover 99% of the recurrent parent genome. This procedure is too time-consuming, particularly in the context of the competitive nature of modern hybrid breeding programmes, where the turnover times for new lines and hybrids are fast.

We describe in brief, three of the successful cases of marker-aided backcrossing intended towards the improvement of qualitative traits.

MAS for Resistance to Soybean Cyst Nematode

Diseaseresistant phenotypes are often simple and oligogenic in nature. Yet, the difficulties in establishing reliable inoculation and scoring methods can challenge even the best plant pathologist or breeder. An example is the resistance to soybean cyst nematode (SCN; *Heterodera glycines*). A widely used phenotypic assay takes five weeks, extensive greenhouse space and 5–10 h of labour for every 100 plant samples processed. The need of the hour for research groups was to develop tightly linked and breederfriendly markers for the major SCN resistance genes. Due to consistent research efforts, microsatellite or SSR markers have become available for the *rhg1* gene, which is primarily responsible for SCN resistance. Currently, the SSR marker *sat309* located only 1–2 cM away from the *rhg1* forms the basis of most public breeding efforts. This marker alone can be extremely effective in screening breeding populations. Genotypic selections with *sat309* were 99% accurate in predicting lines that were susceptible in subsequent greenhouse assays. Recently, high throughput genotyping using Taqman™ probes for a polymorphism linked to SCN resistance gene, *Rhg4* has been standardised, which is discussed in detail later.

MAS in Developing QPM Genotypes

Normal maize protein is inherently deficient in two essential amino acids, namely lysine and tryptophan. QPM refers to maize genotypes in which the *opaque2* gene is introgressed along with endosperm modifiers that provide the hard kernel texture. QPM, thereby, offers twice the amount of lysine and tryptophan with kernel vitreousness comparable to the normal maize. While the favourable *opaque2* allele is recessive in nature, the endosperm modifiers that provide hard kernel texture (instead of the characteristic soft texture conferred by the *opaque2* allele in the absence of modifiers), are polygenic. Introgression of *opaque2* along with endosperm modifiers into elite inbreds is complicated because of three major constraints: (i) each conventional BC generation needs to be selfed to identify the *opaque2* recessive gene, and a minimum of six such BC generations are required to recover satisfactory levels of recurrent parent genome; (ii) besides maintaining the homozygous *opaque2* gene, multiple endosperm modifiers must be selected; and (iii) rigorous biochemical tests are required to ensure enhanced lysine and tryptophan levels in the selected materials in each breeding generation, necessitating significant labour, time and resources. Although conventional breeding procedures have been used successfully to convert commercial lines to QPM forms, the procedure is highly cumbersome and time-consuming.

Scientists at CIMMYT (International Center for Wheat and Maize Improvement), Mexico have successfully developed an innovative combination based on SSR markers for *opaque2* allele and phenotypic selection for kernel vitreousness (through the accumulation of endosperm modifiers) for conversion of normal maize lines into QPM. Three SSR markers, *umc1066*, *phi057* and *phi112*, which are internal repetitive sequences within the *opaque2* gene are being utilised in PCR-based assays to select individuals carrying the *opaque2* gene in successive backcross generations, and hence, the time required is reduced to one-half. In addition, marker-aided background selection can help recover the same level of recurrent parent genome in three generations as would be achieved by six generations of conventional selections. Also, errors due to recombination frequency between the target gene and the flanking marker can be eliminated, since these SSR markers are located within the target gene itself. Apart from this, routine biochemical tests to ensure high lysine and tryptophan levels need not be carried out at each generation since the presence of *opaque2* is confirmed. Thus, MAS based on SSR markers for conversion of normal maize lines into QPM is simple, rapid, accurate, efficient, cost-effective and complementary to existing breeding protocols.

Marker-Aided Oyramiding of Rice Genes

The most effective approach to combat bacterial blight caused by *Xanthomonas oryzae* pv. *oryzae* is the use of resistant varieties. So far, 25 resistance genes have been identified and some of them have been incorporated into modern rice varieties through conventional breeding. Transfer of *Xa4* gene resulted in the development of many bacterial blight-resistant rice varieties, which have played a vital role in sustaining the rice yields in many countries. However, large-scale and long-term cultivation of varieties carrying *Xa4* resulted in a significant shift of the dominant bacterial blight pathogens. Rice varieties with single *Xa4* gene have become susceptible to the bacterial blight disease in many areas in Indonesia, India, China and Philippines. One way to delay such a breakdown in bacterial blight resistance is to pyramid multiple resistance genes into rice varieties. This can be difficult or nearly impossible with conventional breeding approach due to epistasis of genes, particularly when a breeding line already carries a gene, e.g.

Xa21 that shows resistance to most of the bacterial blight races at the time the varieties are being developed. With the conventional approach, a breeding line with *Xa21* only cannot be distinguished from a breeding line with *Xa21* plus some other genes. International Rice Research Institute (IRRI), Philippines and Punjab Agricultural University, Ludhiana have successfully employed MAS to pyramid genes for bacterial blight resistance. All possible combinations of the four resistance genes, viz. *Xa4*, x*a5*, x*a13* and *Xa21* were pyramided using STS (sequence tagged site) markers. The pyramided lines show wider spectrum or higher level of resistance to the bacterial pathogen. Recently, the Government of Indonesia released two new rice varieties, 'Angke' and 'Conde' through MAS, which are resistant to bacterial leaf blight infection.

Rice blast caused by the fungal pathogen, *Magnaporthe grisea* is another important disease throughout the world. The dynamic evolution of the virulence in the blast fungus coupled with the epistatic action of different resistance genes in the host plant, makes breeding for blast resistance a constant challenge. Molecular breeding approach is being widely employed for the improvement of blast resistance in many high-yielding commercial rice cultivars. Three genes, namely *Pi1*, *Piz5* and *Pita* have been pyramided in a susceptible rice variety, CO39 using RFLP and PCR-based markers for durable blast resistance.

MAS FOR IMPROVEMENT OF QUANTITATIVE TRAITS

Most of the traits of agronomic importance are complex and regulated by several genes. Unlike the case of simply inherited traits that are controlled by one or a few major genes, improvement of polygenic traits through MAS is a complex endeavour. The difficulty in manipulating quantitative traits is related to their genetic complexity, mainly the number of genes involved in their expression and interactions among genes (epistasis). Because several genes are involved in expression of a quantitative trait, these genes, in general, have smaller individual effects on the phenotype, and the effect of the individual genes is not easily identifiable. This warrants repetitions of field tests to characterise accurately the effects of QTLs and to evaluate their stability across environments. The evaluation of QTL by environment interaction (Q × E) continues to be a major constraint on the efficiency of MAS. Besides, epistatic interactions among different regions of a genome can induce a skewed evaluation of QTL effects. Also, if the genomic regions involved in the interactions are not incorporated in the selection scheme, they can potentially bias the selection process.

Despite the proliferation of QTL mapping experiments in recent years, a number of constraints have imposed severe limitations on efficient utilisation of QTL mapping information in plant breeding through MAS. Salient among these constraints are: (i) identification of a limited number of major 'players' (QTLs) controlling specific traits; (ii) the notion that QTL identification is required whenever additional germplasm is used; (iii) inadequacies/ experimental deficiencies in QTL analysis leading to either overestimation or underestimation of the number and effects of QTLs; (iv) lack of universally valid QTLmarker associations applicable over different sets of breeding materials; (v) strong QTL—environment interaction; and (vi) difficulty in precisely evaluating epistatic effects.

Increasing the efficiency of MAS for quantitative traits calls for improved field experimentations/designs, robust mathematical models and sound statistical approaches. For instance, with composite interval mapping (CIM), field data from different environments can be integrated into a joint analysis to evaluate the Q × E interactions, thus, enabling identification of stable QTLs across environments. Besides, with a detailed linkage map, CIM allows a precise identification of the QTL in the genome and better identification of linked QTL (in coupling phase) from the same parental line. DNA-based markers also offer the unique advantage of identification and utilisation of favourable QTL from even a phenotypically inferior parental line (in repulsion phase).

An especially powerful approach for enhancing the QTL mapping in tandem with MAS was proposed by Tanksley and Nelson. The advanced backcross-QTL (AB-QTL) analysis involves crosses between elite germplasm and unadapted genotype/wild relatives with favourable genes/QTLs, followed by two generations of backcrossing for developing several hundred sibling lines. These lines, each retaining different genomic segments of the wild relative/unadapted genotype, are then genotyped using DNA markers. The lines become, in effect, a set of near-isogenic lines that individually dissect the effects of potential QTL in the background of the elite parent. At the same time, the BC lines also provide relatively fixed material for the critical step of replicated phenotypic evaluations. The essence of this strategy—uncovering and accessing the desirable alleles from wild relatives or unadapted genotype—demonstrates that QTL mapping can go hand in-hand with MAS rather than as sequential steps.

QTL-marker associations however, need to be unraveled through more intensive research efforts. It would be critical to find out whether extremely tight linkages between marker loci and QTL may lead to highly conserved allele associations. If so, observing marker allele frequency changes in long-term selection experiments or determining markers that explain significant portions of the combining ability variance in diallel or factorial crosses, might reveal universally applicable markers. Another approach would be to conduct QTL analyses in genetically broad-based panmictic populations by means of highly saturated integrated genetic marker maps. For a maximal efficiency of MAS, direct QTL-allele-specific markers (such as STS markers derived from cloned QTL alleles) are needed. Although success in terms of cloning of QTL alleles is highly limited, map-based cloning and candidate gene approaches would increasingly facilitate isolation and characterisation of agronomically important QTLs, particularly due to the rapid progress being made in genome sequencing of various plants. There have been few reports of successful map-based positional cloning of QTLs till recently. QTLs for glucose and fructose content in tomato *(Brix 9-2-5), fruit weight in tomato (fw 2.2), heading* date in rice

(*Hd1*) and (*Hd6*), etc. have been cloned. Cloning the genetic determinants of QTLs is expected to bridge the missing link in our understanding of the relationship between genotype and phenotype.

QUANTITATIVE TRAIT IMPROVEMENT THROUGH MAS

MAS for Improving Heterotic Performance in Maize

The maize lines B73 and Mo17 are not highly productive, but widely used across the world, and represent the two most important 'heterotic groups' in the US maize breeding programmes. Based on a mapping population derived from the B73 × Mo17 cross, QTLs contributing to heterosis for grain yield were mapped on nine of the ten maize chromosomes. Further, mapping studies suggested that two elite inbred lines, Tx303 and Oh43 contained genetic factors that might improve the heterotic response of the B73 × Mo17 single cross hybrid. Six chromosomal segments each in Tx303 and Oh43 were transferred (using MAS) through three backcross generations into the target lines, B73 and Mo17. Two selfing generations followed the third backcross. Based on the initial evaluations of the testcross hybrids, the better-performing modified lines were selected for intercrossing and were designated as 'enhanced' lines. The single cross hybrids derived by crossing the enhanced B73 × enhanced Mo17 exceeded the hybrid 'checks' by 12 to 15%. This study was one of the earliest demonstrations that marker-facilitated backcrossing can be successfully employed to improve complex traits such as grain yield in maize.

Germplasm Enhancement in Tomato using AB-QTL Strategy

The effectiveness of AB-QTL strategy was demonstrated in tomato by Tanksley and coworkers through a series of studies. Genetic enhancement, through AB-QTL strategy, for various traits of agronomic importance, including fruit quality and black mould resistance in tomato, were accomplished using wild relatives such as *Lycopersicon pimpinellifolium*, *L. peruvianum*, *L. hirsutum* and *L. cheesmanii*.

MAS for Drought Dtress Tolerance in Maize

CIMMYT researchers have made considerable efforts during the past three decades to improve pre- and post-flowering drought tolerance in maize. Although significant progress has been achieved for improving drought tolerance in CIMMYT maize germplasm through conventional breeding, the approach is slow and time-consuming. Use of molecular markers and QTL information based on carefully managed replicated tests showed the potential to alleviate the problems associated with inconsistent and unpredictable onset of moisture stress or the confounding effect of other stresses such as heat. The complex trait of drought tolerance was first broken down into simpler components, such as anthesis-silking interval, that are closely associated with drought tolerance. CIMMYT then conducted a series of experiments on QTL analysis and MAS for transfer of drought tolerance to tropical maize, and obtained encouraging results. An integrated strategy of QTL-mapping, MAS and functional genomics is now being explored to further provide genomic information and tools to effectively complement conventional selection for improving drought stress tolerance in maize.

DESIGNING A BC–MAS SCHEME: SOME IMPORTANT CONSIDERATIONS

In a BC scheme, the strategy is to transfer a specific favourable allele at a target locus from a donor line to a recipient line. The use of DNA markers, which permit the genetic dissection of the progeny at each generation, increases the speed of selection process. In general, the BC scheme aims either at complete or partial line conversion. The objective of a complete line conversion is to develop a line that will have exactly the same genetic composition as the recipient line, except at target loci where the presence of homozygous alleles from a donor line is desired. Partial line conversion means that the conversion is complete when a limited proportion of the do nor genome in an individual is found scattered over the genome in addition to the desirable homozygous alleles at the target gene.

Factors considered critical in a BC–MAS scheme include the number of target genes, the distance between the flanking markers and the target gene (2–20 cM), and the number of genotypes selected in each BC generation. Depending upon the objectives, the experimental design for line conversion through BC–MAS needs to be modulated based on the available resources, nature of the germplasm (e.g. agronomic quality and number of lines to be converted) and technical options available at the marker level. Once the number of target genes to be introgressed has been defined, the next step would be to determine the population size that needs to be screened at each generation, giving a target-selectable population size of 50 to 100 genotypes. Once this is defined, one should determine the desirable recombination frequency between the flanking markers and target gene and the number of genotypes selected at each generation based on the objective and the constraints of the experiment.

The number of BC generations required to achieve the introgression can be predicted based on simulations. When resources are limited or introgression from a donor line into a large number of recipient lines is desired, strategies based on BC–MAS at one target locus solely at one advanced BC generation should be considered. Selection in later generations is more effective because the ratio of the standard deviation to the mean of the donor genome contribution increases as the backcrossing proceeds. The backcross procedure can be terminated after four, instead of six, backcross generations even with small population sizes and limited number of marker data points (MDP). Thus, the marker technology can be advantageous even when the resources in a breeding programme are limited. MAS has the potential to reach in BC3 generation, the same level of recurrent parent genome (RPG) as reached in BC7, without the use of markers9. However, large numbers of MDP and more efficient marker systems are required to unlock this potential.

In the above scheme, it is critical to identify the most convenient set of markers for the allele(s) of interest, because the screening of the whole population has to be conducted at least once at the beginning of each BC generation. With the population size running into hundreds or thousands, such screening can be laborious and expensive. However, this can be optimised using an appropriate combination of DNA markers. With the recent advances in molecular technology, particularly single nucleotide polymorphisms (SNPs), a substantial improvement in the capacity to efficiently screen larger populations can be achieved.

EFFICIENCY OF MAS

Computer simulations have provided powerful tools for analysing the design and efficiency of MAS programmes. Three different selection strategies in a markerassisted background selection programme, namely twostep, three-step and four-step selections, were compared by computer simulation in terms of quicker recovery of a large proportion of the RPG9. The simulations were based on maize genetic map ($n = 10$) with markers spaced about 20 cM apart and with the assumption that the target locus could be scored directly either through phenotype or a marker completely linked to the target gene. Major conclusions from this simulation experiment are as follows:

1. The four-stage sampling strategy which includes (a) selecting individuals carrying the target allele; (b) selecting individuals homozygous for recurrent parent genotype at loci flanking the target locus; (c) selecting the individuals homozygous for recurrent parent genotype at the remaining loci on the same chromosome as the target allele; and (d) selecting one individual that is homozygous for the recurrent parent genotype at most loci (across whole genome) among those that remain, is the most efficient procedure in general.
2. With the four-stage sampling strategy and reasonable population size (50–100), one can expect to find BC3 progeny with at least 96% RPG with 90% probability. It would take six generations of traditional backcrossing to reach this stage, besides the risk of a larger probability of linkage drag around the target gene.
3. Increasing the number of markers genotyped at each generation had little effect. Once the threshold of one marker per 20 cM is reached, additional markers (except perhaps around the target locus) would not be required. The frequency of recombination, and not the number of markers, is a more important limiting factor in reducing linkage drag, which suggests that sampling larger populations with fewer markers makes sense than the reverse.

RECENT ADVANCES IN MAS STRATEGIES AND GENOTYPING TECHNIQUES

Single large-scale MAS

In this strategy, selection of parental lines is first carried out among outstanding elite material for the trait to be improved with the best allelic complementarity. By crossing each selected parental line with a tester (elite line lacking the target trait), segregating populations are developed. Genomic regions of interest for each parental line are identified by combining favourable alleles in the segregating populations (e.g. F3 families and RILs). MAS, based on reliable PCR-based markers to fix favourable alleles at target genomic regions, is conducted only once on large segregating populations derived from crosses between the elite lines. The strategy is suggested to offer three distinct advantages: (i) favourable alleles (selected to improve a specific trait) are derived from two or more sources of elite parental lines in a complementary scheme, disregarding the 'recipient/donor' line concept; (ii) plants with fixed favourable alleles at specific genomic regions are selected at an early generation of recombination and no pressure of selection is applied outside the targetted regions. This assures good allelic variability in the rest of the genome for future line development under various conditions and environments; and (iii) the approach is particularly relevant for pyramiding favourable alleles at cloned genes or major QTLs in new germplasm.

Pedigree MAS

This approach40 is especially relevant for crops such as wheat, where pedigrees of elite germplasm are known. Fingerprinting elite wheat materials must be conducted in a set of lines actively used in the breeding programme, and in elite materials to be subsequently released. The data may be combined with the phenotypic data collected during different selection cycles to identify favourable alleles for trait(s) of interest. For example, if an elite line contains alleles for yield performance in a target environment, their frequency should be higher than the expected random frequency in offspring derived from this elite parental line. This shift in allelic frequency reflects phenotypic selection by breeders and may be identified by comparing fingerprinting data of both parents and their offspring. Once the favourable alleles are identified, DNA markers closely linked to the target genomic regions can be used to accelerate fixation of favourable alleles in the next selection step: a new set of elite materials (offspring 1) to derive the next set of elite lines (offspring 2). This MAS strategy was suggested to be most efficient when conducted on F_2 or F_3 segregating populations.

Breeding by Design

'Breeding by design' is a novel concept that aims to control all allelic variations for all genes of agronomic importance. The authors of this concept propose that by understanding the genetic basis of all agronomically important characters and the allelic variation at those loci, the breeder breeder would be able to design superior genotypes *in silico*, which demonstrates that DNA markers not only improve selection processes but can aid in creating novel genotypes bearing new characteristics of agronomic importance. The strategy involves three well-defined steps: (i) mapping loci involved in all agronomically relevant traits, preferably through introgression line (IL) libraries that are extremely powerful not only in mapping, but also in reducing the complexity of polygenic traits by separating them into a set of monogenic loci. An IL library consists of a series of lines harbouring a single homozygous donor segment introgressed into a uniform cultivated background. (ii) Assessment of the allelic variation at those loci through intensive chromosome haplotyping and extensive phenotyping of all agronomic traits. Haplotypes refer to a combination of linked marker alleles that occur in a locus in a set of accessions. (iii) Designing superior genotypes comprising a combination of favourable alleles at all loci through accurate selection of recombination events using flanking markers to collate different favourable alleles next to each other.

NON-GEL BASED SEPARATION OF PCR-AMPLIFIED DNA PRODUCTS

The efficiency of MAS in larger breeding populations could be improved if the laboratory throughput is high and costs are low. Currently, the most common way to evaluate the outcome of PCR amplification is to analyse the amplified products using gel electrophoresis once thermal cycling is complete. This process takes extra time and effort and also increases the chance of laboratory equipment or reagents being contaminated. Consequently, attempts have been made to develop means to detect the accumulation of PCR products either during thermal cycling or following thermal cycling without having to remove the samples for processing. With the advent of third-generation markers, such as SNPs, dispensing with electrophoresis as a core technology to detect polymorphism has become the primary focus. All major SNP detection techniques rely on an initial PCR amplification of the target DNA segment, while they differ in the ways in which the alternate amplicon sequences are discriminated from one another. The SNP is targetted either internally within the amplicon (Taqman™, molecular beacon, etc.) or/ at immediately downstream of the 3¢ end of one of the amplification primers, using allele specific oligonucleotides. The probes employed in both molecular beacon and Taqman™ technologies are based on the principle of fluorescence resonance energy transfer.

Recently, non-electrophoresis-based PCR assays for the allelic discrimination at two linked loci flanking an important QTL controlling days to pollen shed in maize were described. The assays are based on the fluorogenic 5'-nuclease procedure (Taqman), which allows for direct detection of the PCR product by the release of a fluorescent reporter dye as a result of DNA amplification. The main steps in the reaction sequence are polymerisation, strand displacement and cleavage. Two dyes, a fluorescent reporter ® and a quencher (Q) are attached to the fluorogenic probe. When both dyes are attached to the probe, reporter dye emission is quenched. During each extension cycle, the DNA polymerase cleaves the reporter dye from the probe. Once separated from the quencher, the reporter dye fluoresces. The assays were developed after sequencing the alleles at both the loci, by designing suitable primers and probes based on SNPs or insertion/ deletion polymorphisms. The Taqman procedure allowed for a fast and highly reproducible analysis directly in the PCR vials. The complete automation of the process makes this technique highly valuable for the largescale screening required for MAS studies and map-based cloning projects. Similarly, high throughput genotyping using Taqman probes for a polymorphism linked to SCN resistance gene, *Rhg4* has been standardised47. This method can be used for scoring polymorphism in an RIL population and for scoring parental lines in breeding programmes. The Taqman method of determining genotype was accurate in 90% of scores in the RIL population compared to 95% accuracy with electrophoresis. Therefore, this method can be applied to automated large-scale genotyping for soybean breeding programmes.

The molecular beacons have been demonstrated to be useful in discriminating between transgenic and non-transgenic cereals. The hairpin stem formed by the complementary arm sequences of the beacon cannot coexist with the double helix that is formed when the probe hybridizes to its target. Consequently, the molecular beacon undergoes a conformational change that forces the arm sequences apart and causes the fluorophore to move away from the quencher, thereby allowing fluorescence. Oligonucleotide ligation assay and dynamic allelespecific hybridization have been developed, which can potentially increase the power of DNA genotyping. Another interesting and innovative assay is invader SNP assay, which is PCR- and gel-free. Invasive cleavage is a novel non-PCR technique relying on the action of a special endonuclease to remove overlapping oligonucleotides during polymerase-induced extension. These assays are performed on total genomic DNA using universal invader and fluorescent probes and an allele-specific signal probe. The invader assay employs endonucleases (derived from *Archaea* species) from the FEN family (cleavases) to cleave structure-specific rather than sequence-specific sites.

MAS IN PLANT BREEDING: NEED FOR CAUTIOUS OPTIMISM

Integrating molecular marker technologies such as MAS into breeding strategies could become increasingly important in the coming years, to realise genetic gains with greater speed and precision. Although MAS is currently used more widely for simply inherited traits than for polygenic traits, with the development and access to reliable PCRbased markers such as SSRs and SNPs, in several crop plants, efficiency of genotyping large populations or breeding materials has significantly increased. Marker-assisted foreground selection is already gaining rapid momentum as allele-specific markers are now becoming available, through crop genomics research, on a number of agronomically important traits.The promise of MAS for improving polygenic traits in a quick time-frame and in a cost-effective manner is still elusive. There is a wider appreciation that simply demonstrating that a complex trait can be dissected into QTLs and mapped to approximate genomic locations using DNA markers would not serve the ultimate goal of trait improvement. As suggested by Young, research on quantitative traits need to utilise better scoring methods, larger population sizes, multiple replications and environments, appropriate quantitative genetic analysis, various genetic backgrounds and, whenever possible, independent verification through advanced generations or parallel populations. In facing the challenge of improving several lines for quantitative traits, MAS strategies should probably concentrate on using DNA markers in one key selection step to maximise their impact. They could be used at the very beginning of the scheme as predictive tools to reduce the number of crosses, at an early stage of recombination to fix target genomic regions, or at an advanced stage of germplasm development as a diagnostic tool, when the allelic value has already been identified.

The field of 'functional genomics' is making rapid strides, aiming towards assessing the gene functions through genome-wide experimental approaches54. Innovative tools such as DNA chips, microarrays, expressed sequence tags, etc. help in the quantitative assessment of RNA levels in the sample, which in turn might benefit the breeder in selecting best lines based on RNA expression profiles as much as marker genotypes. These innovative approaches are not only expected to identify more genes involved in regulating different pathways in response to stress conditions, but also mark a significant paradigm shift in the field of molecular breeding. In future, quantitative genetics will look towards genomics for information to develop more biologically meaningful models, while genomics will look towards quantitative genetics to develop and validate hypotheses involving complex gene interactions. Bioinformatics will play an important role in facilitating this crossover. Thus, integrating genomics and bioinformatics into the field of molecular breeding is expected to bring in more fundamental revolutions in plant breeding. It would be crucial to recognise the complementarities between molecular technologies and conventional breeding and harness this vital synergy into developing comprehensive research strategies aimed towards more efficient crop improvement in future.

REFERENCES

Ahn, S. N., Bollich, C. N. and Tanksley, S. D., RFLP analysis of genomic regions associated with cooked-kernel elongation in rice. *Theor. Appl. Genet., 1992, 84, 825–828.*

Allard, R. W., *Theor. Principles of Plant Breeding,* Wiley, New York, 1960.

Chu, S. *et al.*, The transcriptonal program of sporulation in budding yeast. *Science,* 1998, 282, 699–705.

Dweikat, I., Zhang, W. and Ohm, H., Development of STS markers linked to hessian fly resistance gene H6 in wheat. *Theor. Appl. Genet., 2002, 105, 766–770.*

El-Kharbotly, A. *et al.*, Segregation analysis and RFLP mapping of the *R1* and *R3* alleles conferring race specific resistance to *Phytophthora infestans* in progenies of dihaploid potato parents. *Mol. Gen. Genet., 1994, 242, 749–754.*

Gilles, G. *et al.*, Single nuclotide polymorphic discrimination by an electronic dot blot assay on semiconductor microchips. *Nature Biotechnol., 1999, 17, 365–370.*

7

Role of Molecular Markers in Plant Breeding

The rapid increases in productivity in food production in recent decades have had improved plant varieties as a basic building block. Those varietal improvements have led to substantial gains in benefits globally. At a more local level, there have been substantial gains in both wheat quality and yield in NSW in recent decades. These varietal improvements have come from the continued efforts of plant breeders to develop varieties that have the characteristics required by farmers to meet the production constraints in the different production environments. Plant breeding programmes have traditionally used "phenotypic selection", combining direct measurement in the field, glasshouse or laboratory with statistical analysis, to allow breeders to search efficiently in large plant populations for progeny that exhibit desired traits. In recent decades, there have been significant technological improvements in both the screening methods available and in the statistical analysis that have allowed that process to become more efficient. At the same time, the development of technologies associated with molecular markers has provided a tool that can replace some of this phenotypic selection and make the breeding process even more efficient. While the new technologies promise significant, often astounding, future benefits, there are several important practical considerations that need to be assessed before the most efficient role for molecular markers in plant breeding programmes can be determined.

MOLECULAR MARKERS

The DNA-based molecular marker is a tool that can identify the presence of particular sections of DNA in an organism. A marker for a trait is defined when a particular segment of DNA is found to be correlated with the phenotypic expression (that is, an observable) characteristic. Some DNA markers can identify the presence of a particular gene ("perfect markers"). Others do not identify a specific gene, but rather a segment of DNA in the vicinity of the genes believed to be associated with the trait of interest ("flanking markers"). For these markers, there is uncertainty about the accuracy of a marker, since their identity is based on correlation, and those correlations are always less than complete. Once a maker has been developed for a particular trait, breeders can use it to accelerate germplasm improvement efforts by:

1. identifying favourable alleles in the genotype (that is, favourable segments of DNA in the breeding line);
2. ensuring that the favourable alleles are carried through into the following generations through repeated cycles of selection; and
3. identifying individual plants in large segregating populations that carry the desirable alleles.

Molecular markers provide a means of rapidly identifying the likely presence of a particular gene of interest (such as a gene for resistance to stem rust) in a breeding line. Marker-assisted selection can thus allow increases in the efficiency of identification and selection of lines with desirable characteristics, which otherwise can take several generations of evaluation. A large number of markers is currently available (with some already being used routinely), and many more are being developed. Key issues identified by Lindner are the intellectual property related to the molecular markers and the access by breeders to those markers. There have been concerns expressed about the use of biotechnology to produce food products, but this aspect of the technology presents no element of environmental or food safety risk. The application of markers discussed here is for the exploitation of genetic variation that exists within the wheat plant or some closely related grass species. These technologies do not lead to alien genes being employed, or changes in genetic structure, but ultimately lead to similar varieties from a breeding programme to those developed in conventional breeding programmes, but may be able to lead to the development of those new varieties in a more efficient manner.

VALUE OF MARKERS TO BREEDERS

Benefits from Application of Molecular Markers

The potential benefits that a breeding programme can obtain from the use of molecular markers are:

1. A significant reduction in the time to develop a new variety. The use of molecular markers can lead to more direct identification of the desirable traits than is possible through phenotypic selection, so that the time to develop a new variety (generally 10 years or more) can be reduced.
2. An increase in the overall rate of genetic gain in a breeding cycle. By allowing the more ready introduction of desirable traits in combination, the varieties produced using molecular markers can be more productive than those developed by conventional breeding methods.

3. The possibility of improving traits that were not possible to measure using traditional phenotypic screening. Some problems have been extremely difficult to assess using phenotypic screening, such as: (i) frost tolerance; (ii) pre-harvest sprouting; (iii) late maturity amylase in wheat; (iv) black point; (v) drought tolerance; (vi) tolerance to exotic pests and diseases such as Russian wheat aphid and Karnal bunt; and (vii) outof-region breeding for constraints not applicable in the region where breeding is taking place.

4. Faster access to new genes to provide greater genetic diversity. By enabling screening for the presence of particular segments of DNA, molecular markers can allow the breeder to identify diversity at an early stage, which can be maintained through the breeding cycle.

While these benefits can potentially be obtained from molecular markers, to date their realisation in wheat breeding programmes has been limited. The reasons for the limited success of practical applications of molecular markers to date include:

1. The relevance of the markers developed to selection criteria of the breeding programme. For example, markers for cereal cyst nematode (CCN) are only likely to provide benefits for breeders developing varieties for the regions where CCN is a production constraint. Similarly, markers for aluminium tolerance are only likely to provide benefits for breeders developing varieties for the acid-soils regions where aluminium toxicity is a production constraint. Where a marker is widely applicable across different programmes and product types, such as markers for protein glutenin composition in wheat, they have been quickly and widely adopted by breeders. Markers relating to specific productivity constraints may be quickly adopted in some programmes while remaining of little interest to other breeders.

2. The extent to which the breeder can readily select for a characteristic phenotypically without a more costly use of markers. Markers for obvious physical traits such as plant height or grain colour can be of little practical interest to a breeder.

3. Even for some more complex traits, there can be a low-cost screening method available to a particular breeding programme that is likely to prove less costly than the application of a molecular marker associated with that trait. For example, a very efficient system of field screening for Septoria tritici blotch has been in place at Wagga for many years, and it is only likely to be replaced by a very low-cost marker.

4. The variation in the extent to which markers are applicable with different breeding populations. While a marker for a particular trait may be very useful to one breeder, it may have no value to another breeder who is using a different breeding population with different genetic variability for the relevant trait. This means that each marker must be validated for each genetic pool before it can be applied as a basis for selection within that pool. Such validation can be time-consuming and expensive for breeding

programmes, given its uncertain outcome, and can lead to a slow rate of adoption of new markers.

5. While some markers can be highly correlated with some characteristics identified by phenotypic screening, that correlation is never complete. Thus, breeders are faced with uncertainty as to the extent to which they can rely on the marker rather than the phenotypic screening. As a result, both systems are often run together in the initial stages of adoption of marker-assisted selection. This can mean an increase in costs during the transition stage between the traditional methods and the use of marker assisted selection.

6. While markers for particular resistance to tolerance genes against pests or diseases can be valuable, the ever-evolving nature of such diseases means that other phenotypic evaluation will still be needed. Therefore, while a marker for a rust-resistant gene can be extremely valuable, relying solely on that marker for selection will make the varieties developed absolutely vulnerable to the break-down of the resistance of that gene, so a breeder would need to still undertake other phenotypic assessment of rust reactions in making selections. Thus, the possible cost savings from the use of that marker can be diluted considerably.

7. The extent to which molecular biologists are not working hand-in-glove with the breeders towards common goals of importance. It can be easier for a molecular biologist to work with more tractable problems, or to find simpler markers, than those that would prove really valuable to a breeder. While this was seen as a significant problem in the earlier years of the development of markers, marker development is currently generally much better coordinated with the needs of breeders. The colocation of the molecular biology laboratory and the breeding programme can be a key factor to assist in making those linkages stronger.

8. Reliable markers have not yet been developed for many characteristics of interest to breeders. For example, there is no "yield" marker, and the nature of the components of high yielding cultivars is such that such a marker may not be feasible, even in the long term.

9. The initial correlation ("linkage") between the marker and the desirable trait may break down or erode over time. Therefore, the breeder may need to periodically conduct phenotypic assessment of the breeding lines to ensure that the marker-assisted selection remains relevant and useful. In the extreme case of a broken linkage, selection for the marker may result in selection *against* the desirable characteristic.

As a result, molecular markers have not replaced conventional evaluation and selection for many traits in the wheat breeding programme at Wagga. Rather, markers are being introduced slowly, on a case-base-case basis, and have at times led to an increase in costs rather than a cost saving during the transition stage when validation is taking place or when the marker is first used.

Value of a Particular Molecular Marker

The value of a molecular marker to a particular breeding programme depends on:

1. costs of the marker implementation
2. costs of alternative phenotypic selection regimes
3. implications for the other operations of the breeding programme.

The value of a molecular marker for a particular trait to an individual breeder will be determined by:

1. the objectives of the breeding programme
2. the relevance of a particular trait in the key target region(s) for the breeding programme
3. the closeness of the linkage between the marker and the desirable trait
4. the capacity to undertake phenotypic screening for the trait by conventional means, and
5. the cost of phenotypic screening for the trait by other means.

As a result, the value of each molecular marker will be different for each breeder. Lindner assumes in his insightful exposition of the provision of essential plant breeding infrastructure that each molecular marker will be ranked in the same order of value by each breeder, but in practice that is unlikely to be the case, even if all breeders were targeting the same set of production environments. For example, some breeders would place different weighting on the production constraints in setting objectives for their programme, so that markers would be given different values. This is magnified where breeders have different target environments. For example, breeders aiming to develop wheat varieties for South Australia are likely to put higher value on markers for boron tolerance and cereal cyst nematode (CCN) resistance than breeders targeting northern NSW, where neither trait is important in the productivity of varieties on farms.

Lindner also assumes in his analysis that each marker will cost the same to develop. Since the key components of marker development involve identification of the molecular marker and its relation to a genetic trait and validation of the linkage between the marker and the trait in relevant breeding populations, the costs are likely to be different. The costs of 6 development of populations for segregating different traits and the costs of the phenotypic evaluations to validate the marker can vary considerably between markers. For example, markers for rust resistance genes can be relatively less expensive to develop, given the availability of segregating populations, the knowledge of the genes themselves, and the ease of screening for such resistance. On the other hand, the costs are likely to be higher to develop and validate a marker for pre-harvest sprouting tolerance, where less is known of the genes involved, segregating populations are more difficult to develop, and the costs of assessing the degree of pre-harvest sprouting tolerance are high. As a result, each marker is likely to have a different development cost, and its validation for different target breeding environments is also likely to vary. The time lags involved will also vary between markers. The value of those markers will in turn vary for breeding programmes that have even subtle differences in objectives.

COSTS OF MARKER-ASSISTED SELECTION

Cost Components

Once a marker has been developed and validated for a given environment and gene pool, it is ready for implementation into the breeding programme, if it is advantageous. The costs of the implementation of a marker in a breeding programme depend on a number of factors:

1. the cost and conditions of the intellectual property relating to that marker;
2. the costs of tissue collection and storage;
3. the method of extraction of DNA/protein for analysis;
4. the extent to which extracted DNA is used for a number of different markers;
5. the analytical method used;
6. the rate of throughput of the analytical system;
7. the degree to which the analysis is combined with that for other markers ("multiplexing");
8. the ease of scoring the alleles after the analysis;
9. the costs of the labour required for the different operations (Dreher *et al.* 2003);
10. the capital cost of the laboratory equipment required;
11. the utilisation of the equipment required (that is, the overall throughput of the laboratory).

The extent to which there are economies of scale in the operations relating to different markers for different crops and situations is not explored. However, such a consideration, along with the extent of the capital equipment required for marker implementation on a routine scale, is important in determining the extent to which these facilities should be centrally located or whether they should be local components of each breeding programme.

Intellectual Property Issues

The intellectual property embodied in a molecular marker is generally protected by patents or similar protective measures, and users are required to pay a licence fee. The form of the fee varies, but is commonly an annual licence fee. In some cases, a one-off fee entitles the breeder to use the marker in an unrestricted way for a number of years. The size of the fees also varies considerably between markers. Where the breeding programme develops its own markers, the fee does not apply, but the development costs would need to be accounted for. The extent to which the marker developed is licensed to other users will have a large bearing on the true cost of the development to the programme.

Costs of Tissue Collection and Storage

In plant breeding programmes, the two forms of tissue generally used for DNA extraction are leaf and grain. In practice, in most cases, leaf samples are used as they are easier to work with and can be obtained within a few weeks of planting rather than waiting for the harvested grain samples. Where grain protein is being assessed, grain samples are required. The leaf samples must be obtained in the field or glasshouse, and frozen for storage until analysed. Special nurseries may be required to develop the plant material for extracting the DNA.

Method of Extraction of DNA/Protein for Analysis

Once the tissue samples have been collected, the next step in applying molecular markers is to extract the DNA from the plant tissue samples. In this operation, a small piece of leaf is placed into a cell (micro-tube) in the rack, along with two small ball-bearings. The rack is then shaken, and the action of the ball-bearings pulverises the leaf sample. A chemical solution is applied to that sample, and the DNA is extracted. Two different methods are commonly used for DNA extraction, using different chemicals (phenol-chloroform and sodium hydroxide). Where protein is extracted from grain samples, a specific protein extraction method is used. Each of these methods has different chemical and consumable costs.

Extent of Usage of Extracted DNA

Since such minute quantities of DNA extract are needed for the analyses, once the DNA has been extracted for marker, it is possible to use the same DNA extract for a number of other markers. This can affect the marginal cost of the different analyses, making some markers economic where they would be more expensive if the full costs of DNA extraction had to be covered by that marker.

Analytical Method Used

There is a wide range of analytical methods that can be used to analyse the DNA for the presence of markers. The specific options available for different markers are determined by the equipment available and the size of the DNA segment required for each of the markers, which can vary substantially. The rate of technological change is very high, and these techniques are evolving rapidly. The chemicals and other consumables differ between methods.

Rate of Throughput of Analytical System

The rate of throughput of the system is a key component of the cost of the application of markers. If each sample were to be run individually, markers would be prohibitively expensive to operate. However, the modern systems are designed to work on arrays of samples, both for the extraction step and for the molecular analysis. The technology now in common usage relies on 96-well arrays undertaking 96 analyses simultaneously. These higher rates of throughput have led to rapid increases in the economic efficiency of the process.

Degree of Multiplexing

"Multiplexing" refers to the concurrent analyses for different markers. Where the markers are located in distinctly separate parts of the genome, analyses can be undertaken simultaneously. This means that one analytical process can enable determination to be made on more than one marker.

Ease of Scoring Alleles after the Analysis

When the analysis has been completed, an operator is required to examine the output or allele scoring to determine the presence or absence of the appropriate DNA band associated with the marker. This is a task requiring considerable skill, and can take a significant amount of time.

Cost of Labour

Since labour is a major component of the cost of the operations, the cost of that labour and the availability of suitably trained operators is a significant factor in the economics of the different operations.

Capital Cost of the Laboratory Equipment Required

The application of molecular markers requires considerable specialised laboratory equipment to undertake the necessary analytical operations. The equipment varies with the different analytical techniques involved. In addition, because of the rapid rate of technological development in the area of biotechnology, the effective life of some of this equipment is shorter than would be expected otherwise. The equipment is likely to be obsolete or redundant long before the equipment physically wears out.

Utilisation of Equipment Required

Clearly, the extent to which expensive specialised laboratory equipment is utilised within the laboratory has a large impact on the costs associated with that equipment.

Empirical Estimates of Costs of Marker Application

As an illustration of the actual costs, empirical estimates are made for two markers currently used in the NSW Department of Primary Industries wheat breeding programme at Wagga Wagga. One marker is associated with resistance to *Septoria tritici* blotch, a leaf disease of wheat, while the other is associated with a low level of polyphenoloxidase enzyme (PPO), a quality characteristic related to Prime Hard wheat. The data used in the empirical analysis are as follows:

1. Intellectual property fee is estimated at $500 per year for the breeding programme.
2. The system uses a 96-well format for the sampling, extraction and analysis.
3. Costs of tissue collection and storage: This operation is undertaken by a Technical Officer who obtains samples for four batches of 96 samples each day.
4. The method used for DNA extraction is the phenol-chloroform method.
5. Two analytical methods were costed, namely the agarose method and the polymerase chain reaction using the genotyper machine (PCR-CEQ) method. For the agarose analysis, one Technical Officer can analyse four 96-sample batches per day, while for the PCR-CEQ analysis only three batches per day are analysed. The chemicals and other consumables are estimated as $0.15 sample for the agarose analysis and $3.68 per sample for the PCR-CEQ analysis.
6. The scoring of the results of the analysis takes 0.5 hours per batch by the molecular biologist for the agarose analysis, while for the PCR-CEQ analysis it requires 1.0 hour of the molecular biologist and 0.25 hours of the Technical Officer per batch.
7. The labour costs are based on salaries and on-costs in the NSW Department of Primary Industries.
8. The estimated total new replacement cost of the equipment in the marker laboratory in Wagga Wagga is approximately $818,000, although only a part of this equipment is used for marker-assisted selection, as some elements are related to research and the development of markers. Annual depreciation and interest charges of approximately $63,000 need to be met from the marker applications analysed. The capital costs per marker vary with the number of each analysis undertaken with each piece of equipment and the costs of the equipment required. The costs per sample range from $1.14 per sample for agarose analysis to $5.21 per sample for PCR-CEQ analysis.
9. No multiplexing is used for these markers.

On this basis, preliminary estimates of the cost components of the two methods are shown in Table 1. The agarose analysis costs $7.97 per sample, while the PCR-CEQ analysis costs $16.33 per sample, on the basis of current throughput levels at Wagga. Labour, operating and capital costs form relatively similar shares of the costs of the agarose analysis, while for the PCR-CEQ analysis the capital costs dominate (and therefore have a greater capacity to fall if throughput levels were increased). However, each component of costs is higher for PCR-CEQ than for the agarose analysis.

Table 1: Estimated Costs of Methods for Marker Application

	Tissue collection and DNA extraction	DNA analysis and scoring	Total
Analysis (Agarose)			
Labour	$1.81	$1.27	$3.08
Operating	$1.35	$1.11	$2.46
Capital costs	$1.28	$1.14	$2.42
Total costs	**$4.44**	**$3.52**	**$7.97**
Analysis (PCR-CEQ)			
Labour	$1.81	$2.07	$3.88
Consumables	$1.35	$4.61	$5.96
Capital costs	$1.28	$5.21	$6.49
Total costs	**$4.44**	**$11.89**	**$16.33**

These costs represent the costs of the application of the methods themselves, and do not include all of the costs of the application of markers. The licence and IP costs, data management costs of the operations and the costs of growing the plant in the glasshouse for the DNA sampling need to be included. The costs of two alternative forms of analysis for applying the marker for Septoria tritici blotch are estimated in Table 2.

Table 2: Estimated Costs of Applying Marker for Septoria Tritici Blotch

	Analysis using Agarose method	Analysis using PCR-CEQ method
Licence and IP costs	$0.22	$0.22
Data management	$0.47	$0.47
Growth of leaf samples	$2.46	$2.46
– *sub-total*	$3.15	$3.15
Tissue collection	$0.91	$0.91
DNA extraction	$3.54	$3.54
Analysis	$3.16	$11.03
Scoring of alleles	$0.37	$0.86
– *sub-total*	$7.97	$16.33
Total costs	**$11.12**	**$19.48**

The preliminary estimates of the total costs per sample of using this marker are $11.12 for the agarose analysis and $19.48 for the PCR-CEQ analysis. The estimated costs of applying the marker for PPO using the same analyses are the same as shown in Table 2, given the same licence fee.

COSTS OF PHENOTYPIC EVALUATION REPLACED BY MOLECULAR MARKERS

Role of Phenotypic Selection Methods

The alternative to the use of molecular markers in a breeding programme is to measure the traits directly in the field, glasshouse or laboratory ("phenotypic assessment") for use as a basis for selection of lines to retain in the programme. For some characteristics such as plant height, this characteristic is readily assessed in the field by a low-skilled operator at a fast rate, and has a low cost. Other assessments need elaborate field sampling and laboratory testing to determine the expression of that characteristic in a breeding population. Other characteristics such as Septoria tritici blotch resistance can be assessed in the field by growing the nursery in conditions to ensure an epidemic, and then scoring the disease expression in the field, a skilled task that can be done rapidly. Root diseases such as cereal cyst nematode and crown rot are much more difficult (and expensive) to score in the field. Glasshouse screening can be used for several characteristics, but can be relatively expensive. A broad characterisation of phenotypic selection for some selected traits is shown in Table 3.

Table 3: Characterisation of Phenotypic Selection for Some Selected Traits

Trait	Where measured	Operator skill	Speed	Expected cost
Plant height	Field observation	Low	High	Very low
Septoria tritici blotch resistance	Field nursery	Medium-high	High	Low
Septoria tritici blotch resistance	Glasshouse	Medium-high	Medium	Medium
Cereal cyst nematode resistance	Field & laboratory	Medium-high	Low	Very high
Crown rot resistance	Field & laboratory	Medium-high	Low	High
Leaf rust resistance	Field nursery	Medium-high	High	Low-medium
Dough glutenins in wheat	Laboratory	Medium-high	Medium	High
Wheat quality (PPO)	Laboratory	Medium	High	Low
Karnal bunt resistance	Overseas (field)	Medium	Very low	High

In this study, the costs of two types of phenotypic assessment were evaluated in detail: (a) Septoria tritici blotch; and (b) wheat quality, measured as polyphenoloxidase (PPO). In each case, the costs were estimated from the operations in place for the wheat breeding programme at the Wagga Wagga Agricultural Institute.

Empirical Analysis of Phenotypic Evaluation of Septoria Tritici Blotch

For Septoria tritici blotch, two forms of evaluation are used, namely field nursery testing and glasshouse screening. The field evaluation takes place under irrigation (centre-pivot) to provide the moist humid conditions favouring a Septoria epidemic. The lines are then scored for the disease during the season by experienced operators, but the nursery is not taken through to harvest. The glasshouse operations are much more labour-intensive. Plants are grown in individual pots, and the inoculum spread to ensure infection. Again the plants are scored for the disease by experienced operators. Because the capacity of the glasshouse is much less than the field (only 300 lines can be evaluated at one time), repeated cycles throughout the season are required in the glasshouse. The costs differ widely between the two evaluations (Table 4). Field testing costs $2.44 per line evaluated, while in the glasshouse the costs are $15.22 per line. In each case, labour is the overwhelmingly major component of the costs, as found by Dreher *et al.* for maize plantings at CIMMYT. While some lines are evaluated only in the field, and some only in the glasshouse, lines at the later stages of the programme are evaluated in both field and glasshouse, which costs a total of $17.66 per sample.

Table 4: Phenotypic Evaluation of Septoria Tritici Blotch at Wagga Wagga

	Field testing	Glasshouse	Total
Labour	$2.14	$14.54	$16.68
Operating	$0.14	$0.15	$0.28
Capital cost	$0.16	$0.53	$0.69
Total costs	$2.44	$15.22	$17.66

Empirical Analysis of Phenotypic Evaluation of PPO

For PPO, the laboratory process is a relatively simple one. A Technical Assistant can complete 504 samples per week, and consumable costs are negligible. Thus, operating costs (including labour) are estimated at $4.96 per line. The equipment needed has annual capital costs of $3,411, but is shared across a number of different analyses. The capital costs amount to $2.21 per sample, or $2.21 per line assessed for PPO. Thus, the total costs of phenotypic assessment of PPO are $7.17 per line (Table 5).

Table 5: Costs of Phenotypic Evaluation of PPO

	Costs per line
Labour	$4.88
Consumables	$0.08
Capital cost	$2.21
Total costs	$7.17

COST COMPARISON OF MARKER-ASSISTED SELECTION AND PHENOTYPIC SELECTION

The broad parameters for the successful implementation of marker-assisted selection are:

1. Marker implementation costs
2. Accuracy of marker

3. Relevance of trait to target environment
4. Capacity for alternative screening
5. Cost for alternative screening

In the case of Septoria tritici blotch, markers cost $11.12 per sample for agarose analysis and $19.48 for PCR-CEQ analysis. Clearly, the former is preferable from a cost point of view. The phenotypic evaluation for Septoria costs $17.66 per line where both field and glasshouse screening is undertaken. Thus, if the agarose-based marker is used, and replaces both field and glasshouse screening, then there can be considerable cost savings. Even replacing only the glasshouse component with the agarose-based marker would also lead to savings. However, for those lines currently only tested in the field screening (the majority of lines), the costs would increase considerably with the application of markers. At present, 7,000 lines are only tested in the field, 1000 are only tested in the glasshouse, and 3,250 are tested in both, which costs $90,000 per year. If this was replaced by agarose-based marker analysis, the total costs would be $125,000 per year (an increase of 40%); if replaced by the PCR-CEQ analysis, the costs would more than double to $219,000 per year. For PPO, the phenotypic evaluation costs $7.17 per line, compared to $11.12 for the agarosebased marker. As a result, replacing the laboratory testing with the marker for PPO would result in a 55% cost increase. If the testing were replaced with a PCR-CEQ -based marker, the costs would almost treble the current cost.

These comparisons are only part of the picture. Where such marker analyses could utilise shared DNA samples, the extraction costs would be reduced, and hence the total costs. For example, if four analyses could be carried out from the one DNA extract, the marker costs would fall by over $5 per sample, or 53% for the agarose analysis and 27% for the PCR-CEQ analysis. Similarly, if the markers could be applied in a multiplexing system, the analysis costs could be reduced, further reducing the total costs of the markers. With three markers multiplexed (*ie* three analysed simultaneously), costs would fall by a further 50%. Thus, the markers would be able to lead to significant cost reductions, where both multiplexing and shared DNA samples were employed. Those gains would, however, only result where the markers were able to fully replace the phenotypic selection. It seems unlikely that such complete replacement will occur in the near future, at least for the Septoria marker, so that those cost savings may not be able to be realised for some time. Only when the markers are sufficiently robust to enable the breeders to replace phenotypic selection altogether for that characteristic, along with an efficient system marker application, are the long-promised cost savings of markers likely to be captured.The marker-assisted selection and conventional phenotypic selection may not be perfect substitutes. In some cases, markers can provide information on allelic composition that is not available through phenotypic assessment. The differences in time for results from the marker-assisted selection and conventional phenotypic selection for some 13 characteristics can mean that the former provides effectively a different output, in terms of its usefulness in selection. On the other hand, field observation for one trait can lead to valuable information about other traits, so that the information from one approach is not directly comparable to that of the other. The two examples discussed cannot be used to make broad generalisations about other markers. As we have seen above, both the costs and the value of markers varies widely for each marker in each breeding programme. As a result, there will be some markers that can lead to significant cost savings for some breeders, while there will be others that will not lead to cost savings. In the interim, while breeders are assessing the relevance and the value of the wide range of new markers that are available, it can be necessary to run both the phenotypic selection and the markers at the same time, and that leads to increases in costs rather than reductions.

In addition, the costs of both markers and phenotypic evaluation vary with the throughput of the systems used, because of economies of scale. For example, if considerably larger numbers of samples were put through either system, the costs per line evaluated are likely to decline. Thus, the PCR-CEQ costs could fall sharply if greater utilisation were made of the genotyper machine used, as the high capital costs per sample would be markedly reduced. Dreher *et al.* show the importance of throughput in unit costs, and similar analysis needs to be undertaken to make similar comparisons in this context. There are also significant developments taking place in the marker technology, such as genome-wide analysis, whereby potentially hundreds of markers can be applied in a single analysis. These systems offer additional applications to breeders other than simple traitmarker associations. However, in the case of trait-marker associations, they offer a potential to alter the costs for each marker, and may enable the relative costs to be reduced considerably. However, such markers need to be relevant to the particular breeding programme. If the system only allows breeders to reduce their phenotypic selection for a small number of characteristics, because so many of those applied relate to characteristics not relevant to that programme, then there may still be no cost savings.

REFERENCES

Brennan, J.P. and Bialowas, A. (2001), *Changes in Characteristics of NSW Wheat Varieties, 1965-1997*. Economic Research Report 8, NSW Agriculture, Wagga Wagga.

Evenson, R.E. and Gollin, D. (eds) (2003a), *Crop Variety Improvement and Its Effect on Productivity*, CABI Publishing, Wallingford.

Hassall & Associates (2002), *Evaluation of the Co-operative Research Centre for Molecular Plant Breeding: Economic Evaluation Component*, Report prepared for the Co-operative Research Centre for Molecular Plant Breeding, in association with EconSearch, Sydney.

Karakousis, A. and Langridge, P. (2003), "A high-throughput plant DNA extraction method for marker analysis", *Plant Molecular Biology Reporter*, 21 95a-95f, March 2003.

Lindner, B. (2004), "Privatised provision of essential plant breeding infrastructure", *Australian Journal of Agricultural and Resource Economics* 48(2), 301-321.

Morris, M., Dreher, K., Ribaut, J.-M. and Khairallah, M. (2003), "Money matters (II): Costs of maize inbred line conversion schemes at CIMMYT using conventional and marker-assisted selection", *Molecular Breeding* 11, 235-247.

8

Inbreeding Depression in Polyploids

Estimates of the frequency of polyploid angiosperm species range from '30–35% to as high as 80%; most estimates are near 50%. Levels of polyploidy may be even higher in pteridophytes, with some estimates of polyploidy in ferns as high as 95%. Polyploids often occupy habitats different from those of their diploid parents, and have been proposed to be superior colonisers to diploids. Furthermore, most crop plants are of polyploid origin, as noted by Stebbins. In contrast, although genome doubling has been reported from other major groups of eukaryotes, it is not nearly as common in these groups as it is in plants. The question often has arisen as to *why* polyploids are so common and so successful, and several possible explanations have been proposed.

Stebbins considered vegetative reproduction and the perennial habit to be important factors promoting the establishment of polyploids, along with an outcrossing mating system to allow for hybridisation (between species, subspecies, races, populations, etc.) in the formation of the polyploid. Perhaps most important to Stebbins was the availability of new ecological niches. Additional hypotheses for the success of polyploids include broader ecological amplitude of the polyploid relative to its diploid parents, better colonising ability, higher selfing rates, and increased heterozygosity. In fact, many aspects of the genetic systems of polyploids may contribute to the success of polyploid plants. These characteristics range from the molecular level to the population level and include increased heterozygosity, reduced inbreeding depression and an associated increase in selfing rates, increased genetic diversity through multiple formations of a polyploid species, genome rearrangements, and ancient polyploidy and gene silencing.

ALLO- VERSUS AUTOPOLYPLOIDY

We will distinguish among types of polyploids by using Stebbins' classification: Allopolyploids are those polyploids that have arisen through the processes of interspecific hybridisation and chromosome doubling (not necessarily in that order), autopolyploids are those polyploids that have arisen from conspecific parents, and segmental allopolyploids are those that have arisen from parents with partially divergent chromosome arrangements such that some chromosomal regions are homologous between the parents and others are homoeologous; segmental allopolyploids will not be considered further in this chapter.

Allopolyploids are characterised by fixed (i.e., nonsegregating) heterozygosity, resulting from the combination of divergent parental genomes; bivalent formation occurs at meiosis, and disomic inheritance operates at each locus. Autopolyploids may exhibit multivalent formation at meiosis and are characterised by polysomic inheritance. Allopolyploids are considered much more prevalent in nature than are autopolyploids, but even a cursory glance at any flora or list of plant chromosome numbers will reveal multiple cytotypes within many species, even though these additional ploidal levels are not typically accorded species status. Thus, autopolyploids in nature likely are much more common than typically is recognised.

Table 1: Genetic variation (mean values) in diploid (2*n*) and tetraploid (4*n*) populations

Species	P		H		A	
	2n	4n	2n	4n	2n	4n
Tolmiea menziesii	0.240	0.408	0.070	0.237	3.0	3.53
Heuchera grossulariifolia	0.238	0.311	0.058	0.159	1.35	1.55
Heuchera micrantha	0.240	0.383	0.074	0.151	1.41	1.64
Dactylis glomerata	0.70	0.80	0.17	0.43	1.51	2.36
Turnera ulmifolia						
var. *elegans*	0.459	0.653	0.11	0.42	2.20	2.56
var. *intermedia*	0.459	0.201	0.11	0.07	2.20	2.00

INCREASED HETEROZYGOSITY

Roose and Gottlieb showed that allotetraploids in *Tragopogon* had fixed heterozygosity at isozyme loci, representing the combination of divergent genomes. In the allotetraploids *Tragopogon mirus* and *Tragopogon miscellus*, 33% and 43%, respectively, of the loci examined were duplicated. These values are typical of the levels of duplicated loci observed in many allotetraploid plants. Of course, this value varies depending on the extent of allozyme divergence between the diploid progenitors: An allotetraploid derivative of two allozymically similar parents would display lower apparent levels of duplicated loci and fixed heterozygosity than would a derivative of more genetically divergent parents. However, even in those cases where there is no apparent allelic divergence between the parental genomes, the chromosomal segment is still duplicated; the possible fates of duplicated genes are reviewed by Wendel. All allopolyploid individuals are essentially heterozygous through nonsegregating, fixed heterozygosity. Populations of autopolyploids are expected to maintain higher levels of heterozygosity than do their diploid progenitors, and these higher heterozygosities can be attributed simply to polysomic inheritance. For example, assuming simple tetrasomic segregation, selfing of a heterozygous autotetraploid of genotype *aabb* is expected to produce progeny in the ratio of 1 *aaaa*:34 heterozygotes (of various genotypes):1 *bbbb*, a huge increase over expectations for a diploid with disomic inheritance (i.e., 1*aa*:2*ab*:1*bb*).

Empirical studies have demonstrated that autotetraploids with tetrasomic inheritance do indeed have higher levels of heterozygosity than do their diploid parents. For example, *Tolmiea menziesii,* which occurs along the Pacific Coast of North America from central California to southeastern Alaska, comprises diploid populations, which are distributed in the southern portion of the range, and tetraploid populations, which occupy the northern portion. Various measures of genetic diversity were compared in natural populations of the two cytotypes.

At seven polymorphic isozyme loci, a substantially larger number of tetraploid individuals was heterozygous as compared with diploid individuals, and comparisons of diploid and autopolyploid populations in other species show the same pattern (Table 1). Many other polyploids also exhibit polysomic inheritance (Table 2); consequently, these polyploids likely also maintain higher levels of heterozygosity than do their diploid parents, simply because of their mode of inheritance.

Table 2: Examples of polysomic inheritance

Species	Inheritance	Evidence
Allium nevii	Tetrasomic	Isozymes
Chrysanthemum morifolium	Hexasomic	Morphology
Daetylis glomerata	Tetrasomic	Isozymes
Dahlia variabilis	Tetrasomic	Morphology
Haplopappus spinulosus	Tetrasomic	Isozymes
Heuchera grossulariifolia	Tetrasomic	Isozymes
Heuchera micrantha	Tetrasomic	Isozymes
Lotus corniculatus	Tetrasomic	Cyanogenic markers, isozymes
Lythrum salicaria	Tetrasomic	Morphology
Maclura pomifera	Tetrasomic	Isozymes
Medicago falcate	Tetrasomic	Morphology, isozymes
Medicago sativa	Tetrasomic	Morphology, isozymes
Pachycereus pringlei	Tetrasomic	Isozymes
Phleum pretense	Hexasomic	Morphology
Solanum tuberosum	Tetrasomic	Morphology, isozymes
Tolmiea menziesii	Tetrasomic	Isozymes
Turnera ulmifolia		
var. *elegans*	Tetrasomic	Isozymes
var. *intermedia*	Tetrasomic	Morphology, isozymes
Vaccinium corymbosum	Tetrasomic	Isozymes

OUTCROSSING RATES IN POLYPLOIDS

Some aspects of polyploid success have been attributed to improved colonising ability, which may involve higher selfing rates than those of the diploid parents. We will present both theory on why we might expect increased selfing in polyploids and empirical data for selected ferns and angiosperms. Theoretical models predict reduced inbreeding depression in polyploids relative to their diploid parents, because of the buffering effect of additional genomes: Deleterious alleles are masked by the extra genomes. Both allopolyploids and autopolyploids are expected to have reduced inbreeding depression, and the magnitude of inbreeding depression is negatively correlated with selfing rates in diploid angiosperms and gymnosperms.

Unfortunately, few studies have addressed levels of inbreeding depression in polyploids empirically; most of the available data come from ferns. Inbreeding depression in ferns has been estimated by taking advantage of the life cycle that involves a free-living, haploid gametophyte generation that can, in most cases, self-fertilise to produce a completely homozygous diploid sporophyte. These studies have involved culturing gametophytes in isolation, in sib pairs, and in non-sib pairs. The number and survival of sporophytes resulting from these treatments are recorded, and these data can be used to estimate inbreeding depression, outcrossing depression, and the number of lethal equivalents per genome. If a greater number of normal sporophytes is produced by non-sib pairs of gametophytes than by either sib pairs or isolated gametophytes, then the population or species is considered to exhibit inbreeding depression.

Masuyama and Watano reported two studies of inbreeding depression in diploid and tetraploid pairs of ferns. In *Phegopteris*, 30–60% of selfed gametophytes of the diploid race formed sporophytes, and nearly 100% of all selfed gametophytes of the tetraploid race formed sporophytes. In *Lepisorus*, only 4% of selfed gametophytes of the diploid race produced normal sporophytes, whereas 98–100% of the gametophytes of the tetraploid race formed sporophytes. These data were interpreted as evidence for reduced inbreeding depression in the tetraploid, with the lower inbreeding depression allowing for increased selfing rates. There are few estimates of selfing rates in polyploid fern species, largely because polyploid fern populations often lack sufficient levels of segregating allozyme markers; however, selfing rates have been estimated in a few diploid-tetraploid pairs.

In *Polystichum*, the allotetraploid *Polystichum californicum* has a selfing rate of 0.236, whereas selfing rates in the two diploid progenitors, *Polystichum dudleyi* and *Polystichum imbricans*, are only 2–3%. In tetraploid *Pteris dispar*, selfing rates are 0.84, much higher than the rate of 0.01 estimated for the diploid race. Limited evidence for ferns suggests reduced inbreeding depression and higher selfing rates in tetraploids than in diploids. Comparisons of outcrossing rates and levels of inbreeding depression in diploid and polyploid angiosperms also are rare. Both outcrossing rates and inbreeding depression have been estimated for diploid and autotetraploid populations of *Epilobium angustifolium*. Outcrossing rates in the two cytotypes were very similar; values (after correcting for inbreeding depression in the diploid) are 0.45 and 0.43 for diploids and tetraploids, respectively. However, the tetraploids have substantially lower inbreeding depression (0.95 for diploids versus 0.67 for tetraploids), as expected from population genetic theory. Outcrossing rates also have been estimated in diploid and allotetraploid species of *Tragopogon*. Outcrossing rates in the allotetraploid *T. mirus* (0.381 and 0.456 for two populations) were higher than those found in the diploid parent *Tragopogon dubius* (0.068 and 0.242), although significantly higher than only one of the two populations; the other parent, *Tragopogon porrifolius*, lacked segregating allozyme variation from which to estimate outcrossing rates.

This pattern is exactly the opposite of that predicted by population genetic theory, and one explanation offered to explain it is that rates of outcrossing were underestimated, particularly in *T. dubius*, because of limited polymorphic loci in all populations. To account for this possibility, outcrossing rates were estimated in *T. mirus* and *T. dubius* from artificial arrays constructed to maximise the chances of detecting an outcrossing event if one had occurred. Outcrossing rates ranged from 0 to .1 for diploid and tetraploid families, and the mean values were quite similar (0.696 and 0.633, respectively, for *T. mirus* and *T. dubius*) and higher than those estimated for natural populations, suggesting that some outcrossing events in both species, and especially the diploid *T. dubius*, had gone undetected. If the outcrossing rates estimated from the artificial arrays are more accurate than are those from natural populations, the discrepancy between predictions and results may be attributable to the recent ancestry of *T. mirus* and to the limited time for the mating systems to have diverged.

RECURRENT POLYPLOID FORMATION

The application of isozyme analysis and DNA techniques to the study of polyploid ancestry dramatically altered our view of polyploid origins. Although morphological or cytological differences among populations of a few polyploid species suggested evidence of repeated polyploid formation, most polyploid species, until recently, were considered to have had a unique origin. Nearly all polyploid species of plants that have been examined with molecular markers have been shown to be polyphyletic, having arisen multiple times from the same diploid species. Polyphyletic polyploid species have been reported for mosses, ferns, and many angiosperms, and include both autopolyploids and allopolyploids. Recurrent formation of a polyploid species has implications for the taxonomy of polyploids, our understanding of the ease with which and rate at which polyploidisation can occur, and, most relevant here, the genetic diversity of polyploid "species." Two allotetraploid species of *Tragopogon*, *T. mirus* and *T. miscellus*, arose within the past century in the Palouse region of eastern Washington and adjacent Idaho from diploid progenitors that had been introduced to the region from Europe in the early 1900s

(Fig. 1). During the past several decades, the ancestries of these two tetraploids have been investigated by using nearly every technique that has become available, and Ownbey's interpretations have been confirmed. Early morphological and cytological data suggested multiple origins of each species, two of *T. miscellus* and three of *T. mirus,* in different locations on the Palouse. Recent isozyme and DNA analyses have supported Ownbey's original hypotheses of recurrent origin and have identified additional lineages of independent formation.

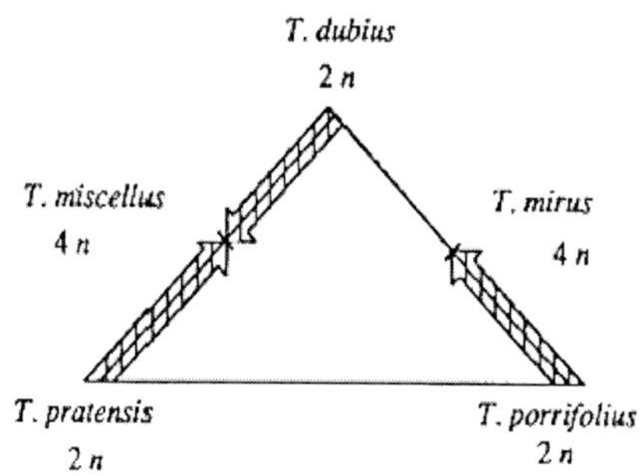

Fig. 1. Parentage and reciprocal origins of tetraploid species of *Tragopogon* in North America.

However, several populations of *T. mirus* in different locations had the same isozyme multilocus genotype, chloroplast DNA haplotype, and rDNA repeat, and, in many cases, they co-occurred with the diploid progenitor species, *T. dubius* and *T. porrifolius*; the same was true of *T. miscellus,* which co-occurred in at least some locations with both of its progenitors, *T. dubius* and *Tragopogon pratensis*. It was possible that these separate locations represented independent sites of polyploid formation from genetically identical (based on the markers at hand) diploids. However, this hypothesis could not be tested without the use of more sensitive markers.

Cook *et al.* used random amplified polymorphic DNA (RAPD) markers to test the hypothesis that isozymically identical populations of *T. mirus* having the same chloroplast DNA haplotype and rDNA repeat were of separate origin and that "identical" populations of *T. miscellus* also were of separate origin. For *T. mirus,* five populations with isozyme multilocus genotype 1 and two populations with isozyme genotype 2 were sampled. Each population had a unique RAPD profile (and, in fact, two populations were polymorphic), suggesting that each population may have had a separate origin. Taken with other data, *T. mirus* may represent a collection of as many as 11 lineages. RAPD data for three populations of isozyme geno-type 1 of *T. miscellus* demonstrated that all three were distinct and possibly of separate origin, raising the number of genetically distinct populations of *T. miscellus* to five.

The *Tragopogon* tetraploids represent remarkable cases of recurrent formation on a small geographic scale and in a short period, perhaps the last 70 years. Other polyploid species, if examined in sufficient detail, may be similarly grossly polyphyletic. Furthermore, recurrent formation of *T. mirus* and *T. miscellus* also has occurred on a broader geographic scale. Both species have been reported from Flagstaff, AZ , and *T. miscellus* has been reported from Gardiner, MT, and Sheridan, WY. Although such polyphyly calls into question the meaning of the term"polyploid species," the biological implications of recurrent polyploidisations from the same diploid progenitor species are indeed intriguing. Such multiple formations may play a significant role in shaping the genetic structure of polyploid species, as they are currently recognised. The concept of recurrent formations forces us to consider polyploid species not as genetically uniform, as previous models of polyploid formation imply, but as genetically variable.

GENOME REARRANGEMENTS IN POLYPLOIDS

Another possible source of genetic novelty in polyploids is genome rearrangements. Evidence for chromosomal changes has been obtained through a number of techniques, including genome *in situ* hybridisation (GISH), analysis of restriction fragment length polymorphism (RFLP) loci, and chromosome mapping. Among the earliest studies reporting widespread genomic changes in tetraploids relative to their diploid progenitors is an analysis of tobacco genome structure using GISH. Tobacco (*Nicotiana tabacum*) is an allotetraploid whose parents are *Nicotiana sylvestris* and a T-genome diploid. GISH clearly revealed numerous chromosomal rearrangements. In fact, nine intergenomic translocations have occurred within the genome of tobacco, that is, translocations between the chromosomes donated by *N. sylvestris* and the T-genome parent. Most of the chromosomes of tobacco are therefore mosaics, composed of regions of both parental chromosome sets.

In *Brassica*, there is evidence that such genome rearrangements may occur very soon after the formation of the tetraploid. Song *et al.* produced artificial tetraploids resulting from interspecific crosses between *Brassica rapa* and *Brassica nigra* and between *B. rapa* and *Brassica oleracea*. They compared genome structure in the F5 derivatives of these crosses with their F2 ancestors and found genetic divergence in these few generations, with distances as high as almost 10%. In addition, Song *et al.* found evidence of cytoplasmic–nuclear interactions—the maternal genotype had definite control over aspects of the nuclear genome. They concluded that a possible result of polyploid formation is the production of novel genotypes. Furthermore, extensive genetic change can occur in the early generations after polyploid formation and may therefore be important in the formation of a functional polyploid. Chromosome mapping of diploid *Brassica* and comparison with the map of *Arabidopsis thaliana* suggest that the diploid species of *Brassica* may actually be ancient hexaploids. Such intergenomic translocations are not limited to tobacco and *Brassica*. Instead, extensive chromosomal changes have been reported in a number of other polyploids, including maize, oats, and soybeans. Such intergenomic translocations may be mediated by transposable elements and may be an important source of genetic novelty in polyploids.

Table 3: Angiosperm families with high chromosome numbers

Basal angiosperms	Chromosome number, *n*
Family	
Illiciaceae	14
Schisandraceae	14
Lauraceae	12
Calycanthaceae	12
Magnoliaceae	19
Eudicot families	
Trochodendraceae	19
Platanaceae	21
Cercidiphyllaceae	19
Salicaceae	19
Hippocastanaceae	19

ANCIENT POLYPLOIDY AND GENE SILENCING

Basal Angiosperms

Estimates of ancient polyploidy generally have relied on chromosome number alone; Stebbins, for example, viewed those plants with a base chromosome number of *n* 5 12 or higher to be polyploid, and others used similar criteria. Based on this criterion, a large number of angiosperm families, most of which trace their roots far back into angiosperm phylogeny, are considered to be the products of ancient polyploid events whose diploid ancestors are now extinct. For example, the Illiciales have *n* 5 14, and both the Lauraceae and Calycanthaceae of Laurales have a base number of *n* 5 12. The lowest chromosome number in the Magnoliaceae is *n* 5 19, and the family exhibits a range of numbers that are multiples of this base number.

Some early eudicots, such as *Trochodendron* and *Tetracentron* (with *n* 5 19) and *Platanus* (with *n* 5 21), also have high chromosome numbers. Some families of more recent origin also are considered ancient polyploids. Some families of possible ancient polyploid origin, along with their chromosome numbers, are listed in Table 3, and the phylogenetic distribution of these families is shown in Fig. 2. Stebbins also suggested that the ancestral base chromosome number for angiosperms is *x* 5 6, 7, or 8. Reconstruction of chromosomal evolution across the angiosperms is partially consistent with Stebbins' hypothesis. Although the high chromosome numbers of the basal angiosperm groups make it difficult to infer base chromosome numbers for those groups of angiosperms and therefore for angiosperms as a whole, our reconstructions show an ancestral number of *x*58 for the eudicots (D.E.S., unpublished data), that is, the large clade that makes up 75% of angiosperm species.

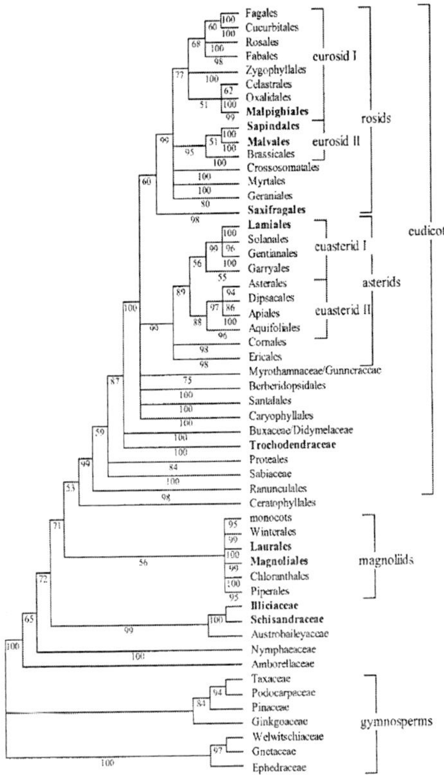

Fig. 2. phylogenetic tree of angiosperms

Most, if not all, angiosperms may have experienced one or more cycles of genome doubling, and these hypotheses of ancient polyploidy have several implications for the genetics, genomics, and evolutionary biology of these plants. First, if they are indeed polyploids, then these plants should exhibit extra copies of their genes above the level that one would expect for diploid plants. Analyses of enzyme expression indicate that multiple enzymes are indeed expressed in putatively paleopolyploid angiosperm families, such as those listed in Table 3; issues of the regulation of duplicated genes are discussed by Wendel. Second, some copies of these multiple genes might be expected to be silenced, particularly in the more ancient families. Third, reorganisation of the original polyploid genome might have led to a novel genomic arrangement and perhaps to novel phenotypes. Finally, given that all members of a family have chromosome numbers that are multiples of a single lower number, it appears that, after polyploidisation, diversification continued at the new polyploid level, with subsequent episodes of polyploidy superimposed on this initial polyploid level.

Homosporous Pteridophytes

Homosporous pteridophytes are those ferns, lycophytes, and *Equisetum* with a homosporous life cycle; all of these groups are the descendants of ancient plant lineages that extend back to the Devonian Period. The mean gametic chromosome number for homosporous pteridophytes is *n* 5 57; for angiosperms, it is *n* 5 16. Despite their high chromosome numbers, however, homosporous pteridophytes exhibit diploid gene expression at isozyme loci. At least two possible explanations can explain this paradox of high chromosome numbers and genetic diploidy. First, these plants are ancient polyploids that have undergone extensive gene silencing to produce genetic diploids, and second, they may have achieved high chromosome numbers through another mechanism, such as chromosomal fission.

Gene Silencing

Genes duplicated through polyploidy have several possible fates: retention of both copies as functional genes, acquisition of new function by one copy, and gene silencing. Several models of genome evolution, in which a polyploid genome gradually will undergo gene silencing and return to a diploid condition, have been presented. Unfortunately, little empirical evidence is available to support or to refute these models. Potential examples occur in the homosporous pteridophytes. Data for the ferns *Polystichum munitum* (*n* 5 41) and *Ceratopteris richardii* (*n* 5 39) may address these alternatives. Pichersky *et al.* studied the genes for the chlorophyll *ayb* binding proteins in *P. munitum.* These proteins are important in photosynthesis and are encoded by a small multigene family. *P. munitum* exhibits diploid isozyme expression. If this species is of ancient polyploid origin but has since undergone substantial gene silencing, then pseudogenes should be detectable in the genome. Five clones of the *CAB* genes were analyzed by Pichersky *et al.* Three of the five clones were structurally nonfunctional, a fourth clone had a structurally intact sequence but was nonfunctional at the sequence level, and a fifth clone was a functional sequence. Possible explanations for these results are (*i*) amplification of nonfunctional sequences in the genome of *P. munitum,* regardless of the ploidy of *P. munitum,* (*ii*) *P. munitum* is diploid with a large number of mutant *CAB* genes, and (*iii*) *P. munitum* is polyploid, with silencing of multiple genes that are present because of ancient polyploidy.

REFERENCES

Grant, V. (1981) *Plant Speciation* (Columbia Univ. Press, New York), 2nd Ed.

Masterson, J. (1994) *Science* 264, 421–423.

Stebbins, G. L. (1950) Variation and Evolution in Plants (Columbia Univ. Press, New York).

Wendel, J. F. (2000) *Plant Mol. Biol.* 42, 225–249. Hickman, J. C., ed. (1993) *The Jepson Manual* (Univ. of California Press, Berkeley).

9

Potato Breeding

One hundred years of advertising by the seed potato industry have convinced farmers that they should buy new seed tubers for every potato crop. This is in order to control tuber-borne *diseases,* and particularly *virus* diseases. And a comparable period of advertising by the chemical industry has convinced farmers, to say nothing of crop scientists, that the best way to control potato pests and diseases is to spray every potato crop, repeatedly, with various crop protection chemicals. It is in the interests of both these industries that modern potato cultivars should be susceptible, so that farmers *need* certified seed, and they *need* crop protection chemicals.

PLANT BREEDING

An individual can easily go it alone, particularly if he (or she) is a small farmer or market gardener. And the rewards for a successful cultivar can be huge. However, a group of like-minded individuals who organise themselves into a plant breeding club will have many benefits, because they can share their enthusiasm, their knowledge, their work, their costs, their sense of achievement, and their royalties. Possibly the most effective club is the university breeding club, operated by students under the supervision of a professor. An amateur breeding club may wish to associate itself with a university breeding club, for their mutual cooperation and benefit. If there were 100 potato breeding clubs in the world, each screening 100,000 true seedlings a year, there would be a total of 10 million seedlings being screened annually. After one or two decades of cumulative breeding, there would be:

1. A virtual elimination of potato parasites, worldwide,
2. A virtual elimination of the need for crop protection chemicals in potato crops, worldwide,
3. An elimination of the recurrent need for expensive certified seed, and
4. Several near-perfect, but different, potato cultivars for each agro-ecosystem.

There seems to be *little doubt* that these targets can be achieved with horizontal resistance, and this is a possibility that clearly merits investigation.

Most universities have photography clubs, chess clubs, and many others. So there is no reason why they should not have plant breeding clubs also. Some agricultural universities may care to convert their breeding clubs into official teaching courses. The students would do all the work of plant breeding, for a full *breeding cycle*, and this would earn them course credits. The professor in charge would earn teaching credits. Graduates would be given life membership in their clubs and they could both benefit from new club cultivars, and they might contribute breeding material of their own. Graduates would also be encouraged to start amateur breeding clubs in their place of work. University clubs might also link to amateur clubs, or even secondary school clubs, to their mutual benefit.

It is a fundamental principle of ecology that diversity leads to stability. Conversely, uniformity leads to instability. Consider the uniformity of a huge area of wheat, all with the same vertical *resistance*. The failure of that resistance to a new race of a parasite represents a major instability. One of the advantages of having many plant breeding clubs, each producing new cultivars for its own agro-ecosystem, is that there would also be a great diversity of cultivars, and a great stability of agricultural production.

POTATO PLANT

The modern, cultivated potato (*Solanum tuberosum*) is a tetraploid. That is, it has four sets of chromosomes, instead of the usual two sets that occur in the more normal diploid plants. This means that the cultivated potato cannot easily be crossed with many wild potatoes, which are mostly diploids. Consequently, amateur breeders should not attempt inter-specific crosses. All their potato breeding should be conducted within the cultivated species, but this is not a serious limitation because this species exhibits immense genetic variation. The potato tuber is the swollen end of an underground stem called a stolon. The number, shape, and size of the tubers of one plant are variable, but the variation between cultivars is even greater.

Potatoes are propagated vegetatively, as clones, using 'seed' tubers. Except for an occasional mutation, or 'sport', clones do not exhibit genetic variation. This is a considerable advantage for both breeders and farmers, as a true seedling can be immediately stabilised genetically, simply by producing a clone. Potatoes produce flowers that can be either self-pollinated, or cross-pollinated, to produce fruits and true seed. This true seed exhibits great variation, and it is the basis of potato breeding.

Any mechanism that protects a plant from its parasites is either stable or unstable. A stable mechanism is unaffected by genetic changes in the parasite. An unstable mechanism is liable to stop functioning when the parasite produces a new strain that is immune to that mechanism. DDT-resistant houseflies and mosquitoes are a classic example of an unstable protection. Stable insecticides include natural pyrethrins and rotenones, soft soap, and oil films on water that kill mosquito larvae. These insecticides never break down to new strains of the parasite. Stable fungicides include those made from copper and the dithiocarbamates.

Figure 1. Potato Plant

Most modern synthetic crop protection chemicals are unstable, and they stop functioning when new strains of the parasite appear. This distinction between stable and unstable also applies to host resistance to parasites. Stable resistance, which is *beyond* the capacity for genetic change of the parasite, is called horizontal resistance. And unstable resistance, which is *within* the capacity for genetic change of the parasite, is called vertical resistance.

HORIZONTAL RESISTANCE

Horizontal resistance is stable resistance. It is durable resistance. It will never break down to new strains of the parasite. Justification for this contention will be found in *Return to Resistance*. The inheritance of horizontal resistance is controlled by many genes of small effect. These genes are called polygenes. The inheritance of horizontal resistance is thus quantitative.

That is, it can be inherited with any degree of difference between a maximum and a minimum. The effects of horizontal resistance are also quantitative. That is, the resistance may be exhibited at any level between a maximum and a minimum. In the absence of crop protection chemicals, the minimum level of horizontal resistance normally results in a total loss of crop, while the maximum level results in a negligible loss of crop.

Horizontal resistance to one species of parasite does not normally confer resistance to other species of parasite. When breeding for horizontal resistance, it is necessary to accumulate high levels of resistance to each locally important species of parasite. The accumulation of all these resistances must be done more or less simultaneously. The mechanisms of horizontal resistance are many, and they are mostly obscure. For the amateur breeder, these mechanisms are unimportant.

All that matters is that the level of resistance is increased, and the level of parasitism is reduced. Because horizontal resistance is quantitative and durable, breeding for this kind of resistance is cumulative and progressive. That is, a good horizontally resistant cultivar need never be replaced, except with a better cultivar. This progression of better and better cultivars can continue, with diminishing returns, until a final limit to improvement is reached.

VERTICAL RESISTANCE

Vertical resistance is single-gene resistance, and it was the resistance of choice among professional plant breeders for most of the twentieth century. It has the advantages of conferring a complete protection, and of functioning over a wide geographic area. It has two grave disadvantages, however. It is liable to break down to new strains of the parasite, and it is thus unstable, or temporary, resistance. And it does not occur against many species of potato parasite. Vertical resistance can be a nuisance when breeding for horizontal resistance, because the vertical resistance masks the effects of the horizontal resistance entirely. If this problem is serious, the one-pathotype technique should be used. Alternatively, only parents known to lack these vertical resistances should be used in the breeding program.

In general, this is not a difficult problem in potato breeding because vertical resistances are relatively rare. They occur against *blight*, but the *A2 mating type* effectively eliminates this problem. Vertical resistances also occur against wart disease (*Synchytrium endobioticum*) and golden nematode (*Globodera rostochiensis*), which are absent from most potato-growing areas. They also occur against some of the temperate viruses. These vertical resistances function by means of a mechanism called hypersensitivity. When the microscopic pathogen penetrates a cell, all the surrounding cells react with extreme sensitivity, and they die. The pathogen cannot survive in dead cells and it dies with them. This hypersensitive reaction produces a brown, necrotic 'fleck' that is just visible to the naked eye.

NO GOOD SOURCE OF RESISTANCE

When breeding for vertical resistance, you must first find a good source of resistance. This has become a shibboleth among plant breeders, to the point that they generally believe also that you cannot breed for horizontal resistance without first finding a good source of resistance. This misconception is very important. Breeding for a single-gene resistance must start with that single gene (the 'good source') and, without it, the breeding cannot even begin. But this is *not* true of horizontal resistance, which is polygenically controlled. Start your horizontal resistance breeding with a group of parents that are susceptible, modern cultivars. All that matters is that a reasonably wide genetic base is used. All the polygenes will be present in the breeding population as a whole, even if they occur at too low a frequency in each individual. The breeding technique will increase the polygene frequency using *recurrent mass selection* and *transgressive segregation*. In general terms, it is much easier to increase the levels of horizontal resistance to various locally important parasites, than it is increase other variables, such as the yield and the quality of crop product. For this reason, it is better to start with modern cultivars that have high yields and quality, but low resistance, than it is to start with primitive cultivars, or even wild progenitors, that have high resistances but low yields and quality.

VERTIFOLIA EFFECT

The 'vertifolia effect' is a loss of horizontal resistance that occurs when the plant breeding is conducted either in the presence of a functioning vertical resistance, or under the protection of a fungicide or insecticide. It happens because the level of horizontal resistance can be assessed only by the level of parasitism, and this is possible only if the parasitism is present and active. Individual plants with a relatively high level of horizontal resistance are in a minority. If the level of resistance is invisible because there is no parasitism at all, susceptible individuals, which are in the majority, stand a better chance of being selected on the basis of their other attributes.

This loss of horizontal resistance to potato *blight* began soon after the discovery of the first fungicide, called Bordeaux *mixture*, in 1882. This fungicide made the breeding much easier because so many seedlings were no longer being killed by blight. The gradual loss of resistance was not recognised at the time and, even if it had been, no one would have cared, because it was believed that resistance was not necessary, as the crops could now be sprayed with Bordeaux mixture. All potato cultivars bred after about 1885 were increasingly susceptible. At about this time, in North America, farmers began to spray against Colorado beetle. The only insecticides they had were very nasty compounds of lead, arsenic, cyanide, and mercury. Later, with the use of vertical resistance, the vertifolia effect continued. And then DDT was discovered, as well as many other synthetic insecticides and fungicides. Because of the vertifolia effect, which has now been operating for a century or more, modern cultivars of potato have less horizontal resistance than the cultivars of 1880. Our task is to ensure that they have much more horizontal resistance than those old varieties.

BREEDING CYCLE

A plant breeding cycle would be called a 'generation' in people or animals. A plant breeding cycle begins with pollination, and the subsequent production of true seed. It ends when the next pollination occurs among selected plants that become the parents of the next breeding cycle. Most breeding programs require some 10-15 breeding cycles. In the temperate regions, there will normally be one breeding cycle each year but, with potatoes, it may be possible to complete two breeding cycles each year.

TRANSGRESSIVE SEGREGATION

This term means that progeny may have a higher level of an inherited, quantitative variable than either of their parents. Breeding for horizontal resistance depends very heavily on this phenomenon. Suppose there is a population of wild potatoes in which every individual plant possesses only 10% of the polygenes contributing to horizontal resistance. The population as a whole will be highly susceptible, and so will every individual within that population. But suppose also the every individual possesses a different combination polygenes in its 10% share. This means that all the polygenes are present in the population, but no individual possesses enough of them to be resistant. With random cross-pollination, some of the progeny will have more than 10% of those polygenes (and others will have less). And, given adequate *selection pressures,* this percentage will increase with each successive generation. After enough generations of cross-pollination, there will be progenies that have most, possibly all, of the polygenes. These individuals will be highly resistant. And their resistance will be durable. This breeding process of artificially increasing a variable character by transgressive segregation is called recurrent mass *selection.*

RECURRENT MASS SELECTION

The main feature of recurrent mass selection is that it deals with *quantitative* variables, such as horizontal resistances. And it raises the level of many desired variables simultaneously, including all the horizontal resistances to all the locally important parasites. The method of recurrent mass selection is to grow a large screening population of seedlings which all genetically different. The best 10-20 individuals in that population are selected, and they become the parents of the next screening generation. Each generation (or breeding cycle) exhibits transgressive segregation, and the desired variables increase accordingly. This increase is usually 10-20% in the early breeding cycles, and the rate of increase gradually declines until no further progress is possible. This 'ceiling' is normally reached after 10-15 breeding cycles and, with horizontal resistances, it should provide a complete control of all locally important parasites without any use of crop protection chemicals.

SELECTION PRESSURES

The term 'selection pressure' is used in the sense of bringing pressure to bear, and it applies to variable populations. For example, if a heterogeneous population of potato seedlings is exposed to a parasite such as blight, the most susceptible individuals will be killed, and the least susceptible individuals will survive, and they will reproduce the most. As a consequence, the next generation will be more resistant. The selection pressure is exerted by the blight fungus, and this can happen in both a wild population, and the screening *population* of a breeding program.

CHOOSING THE PARENTS

The breeding should begin with 10-20 parents, and each parent should be a popular modern cultivar. In choosing the parents, the aim is to have as wide a genetic base as possible, while maintaining high standards of yield and quality. There is no need to choose parents on the basis of their resistance to parasites, because a reasonably wide genetic base will accumulate all the horizontal resistance we need by transgressive segregation. If it does not, the genetic base can always be widened later. Because of possible problems in *flower production*, or with *pollen sterility*, it is advisable to use a few more parents than are needed, so that some can be discarded if they prove difficult to cross. Alternatively, new parents can be added to the breeding population at a later date.

NEO-TUBEROSUM

N.W. Simmonds did an interesting experiment in Scotland. He wanted to prove that *Solanum tuberosum* was derived from the South American species *Solanum andigena*. He also wanted to show that horizontal resistance to blight could be accumulated in the very susceptible *S. andigena*. He proved both these points by recurrent mass selection, and he produced a new kind of potato which he called *neo-tuberosum*. This new potato is closely similar to cultivated potatoes, and it has a useful level of horizontal resistance to blight. It has been used in commercial potato breeding, and these new clones would be useful parents for amateur breeders, particularly if they were trying to widen the genetic base.

GRAFTING

Recurrent mass selection requires very large numbers of seedlings, almost without limit. Grafting potato scions on to tomato rootstocks is a technique for producing very large numbers of true seeds of potato. Selected potato scions can be grafted on to tomato rootstocks to induce flowering. Being unable to form tubers, the potato scion grows upwards as a vine, with a bunch of flowers every few inches. These flowers can be used for controlled cross-pollination. One good graft can produce up to fifty fruits, with up to 300 seeds per fruit, totalling some 15,000 true seeds.

There are several steps in this grafting. Grow one tomato plant in each pot so that it is about six weeks old when the grafting is to take place. Each pot should be fairly large (about 12 inch diameter) to avoid any need for re-potting later. With a new razor blade, cut off the tomato stem with a horizontal cut, about 1½ inches above the first two leaves. Then make a vertical cut down the centre of the stem to a depth of about one inch. Cut a vegetative shoot off the potato parent, several inches long. Place it in water and take it immediately to the grafting house. Cut off a terminal shoot about one inch long for use as a potato scion. Slice the sides of the scion to make a wedge. Insert the wedge into the vertical cut in the tomato stem.

Wrap the graft union in a ribbon of very thin, transparent, plastic film. This ribbon serves two purposes. It binds the graft-union surfaces together, and it prevents the union from drying out. It should be fairly tight, but not too tight. Place the entire pot inside a *mist propagator* for about five days, or until the potato scion shows signs of active growth. Alternatively, put three sticks in the pot, and cover the entire pot with a plastic bag that is tied to the top of the pot, in order to maintain a high humidity. The two tomato axillary buds may start growing. If they do, cut them off. The tomato leaves should be removed later, when they show signs of senescence. The ribbon of plastic film may be removed soon after the scion starts growing and the graft union is well calloused over.

Place the pot in a greenhouse with a supporting string hanging from the roof. The potato vine may grow to a height of several feet. A few, small aerial tubers may form, and they should be removed. There should be two grafts of each potato parent, mainly as an insurance in case one graft is lost for any reason. With 10-20 parents, there will be 20-40 grafts. With practice, *cross-pollination* among this many plants should not require more than 1-2 person-hours of work each day.

Tomato rootstocks are useful because we know that they graft easily and well. However, they have the disadvantage that they can get blight (*Phytophthora infestans*) and, if they do, the entire graft is in grave danger of being lost. There is also a risk of blight attacking the potato scion, particularly in the early breeding cycles. There are two ways of avoiding blight on both the stock and the scion. One is to spray them with a copper or dithiocarbamate fungicide. An alternative is to water the plants from the base of the pots, and to keep both the scion and the above-ground stock dry at all times. This is easily achieved in a greenhouse. Blight spores need free water to germinate and, if all the aerial parts of the grafted plant never get wet, they will not get blight. For complete safety, both methods may be used.

Some useful research, that can easily be undertaken by amateur breeders, would involve a search for alternative rootstocks that are immune to blight. Possible species include the common weed called 'thorn-apple' (*Datura stramonium*); wild species of *Solanum*, such as *S. nigrum* and *S. dulcamara* in Europe, or *S. rostratum*, known as prickly potato, or the buffalo burr, in North America; eggplant (*Solanum melongena*); and tobacco (*Nicotiana tabacum*).

FLOWER PRODUCTION

The potato scion will start producing flowers quite soon after grafting. Each inflorescence usually has 5-10 flowers, and they normally open one or two at a time each day.

CROSS POLLINATION

Cross pollination is done in order to produce genetic variability and transgressive segregation. There are two steps in cross-pollination.

1. *Emasculation* involves the removal of the anthers so that the flower cannot self-pollinate. Emasculation is done the day before the flower opens, when the anthers are still immature and infertile. This stage is easily recognised because the petals are fully formed but still adhering to each other, so that the flower is closed. The petals must be gently separated with a needle and then each of the five anthers must be bent away from the pistil until it breaks off and falls to the ground . The next day, an emasculated flower is wide open and is easily recognised by the absence of yellow anthers . Each day, all the flowers that are ready should be emasculated, except those of the male parent.
2. *Cross-pollination* involves the placing of pollen on the now receptive stigma of each emasculated flower. It is a good idea to use only one male parent, and a different male parent, each day in rotation, so that each parent is represented equally. The male parent flowers are wide open and are easily recognised by the presence of yellow anthers. One anther is picked with a pair of fine forceps , and it is touched to the stigma of each of several emasculated flowers . Make sure that the side with free pollen is used. A small dot of yellow pollen should be visible on the stigma. There are five anthers to each male parent flower and, with several flowers, this should provide adequate pollen for all the cross-pollinations to be done in one day.

Some potato cultivars are pollen-sterile, or even female sterile. If it is discovered that a male or female parent never forms fruits, its use must be discontinued. Famous cultivars that suffer sterility problems are *Russet Burbank*, *Bintje* and *King Edward*, and the use of these as parents should be avoided by amateur breeders.

POTATO FRUITS

The true fruits of potatoes are similar to a small tomato. They develop from a pollinated flower. When ripe, they are usually a green or bronze colour, and they are soft and somewhat wrinkled. Each fruit contains up to 300 true seeds. Because of the cross-pollination, these seeds will exhibit great variation in all their quantitatively inherited characteristics, including their horizontal resistances.

SEED PREPARATION

When the potato fruits are ripe, they should be put in a kitchen blender or food processor and covered with clean water. The blender should be switched on intermittently (the 'pulse' button) so that all the fruits are broken open, and seeds liberated. Do not over-do this maceration as the seeds themselves might be damaged. If a blender is not available, crush the fruits and then shake them vigorously in water in a sealed jar. Now pour all the contents of the blender (or jar) into a clean plastic bowl or bucket. Leave the mixture in a warm room to ferment for about twenty four hours. Most of the fruit debris will float, and most of the seeds will sink. The debris can be removed with a coarse sieve, such as a kitchen colander, and put into another bucket. Decant the water from the first bucket into the second bucket. The seeds in the first bucket can now be put into a glass jug which is repeatedly filled with water and decanted, until the seeds are clean and free from rubbish. The seeds can now be put into a conical, paper, coffee filter, in a kitchen sieve, to drain. When drained, the seeds should be spread out on a clean surface, such as glass, to dry. Potato seed often shows dormancy, which can be a nuisance in a breeding program. This process of fermentation breaks the dormancy, and all the seeds will then germinate immediately. They can either be used as soon as they are dry, or they can be stored for future use.

SOIL SIEVES

A set of soil sieves, designed for the mechanical analysis of soil, is a useful device for a final cleaning of dry seeds. The coarsest sieve is at the top, and finest at the bottom. The seed is shaken through the sieves, and the coarse rubbish is trapped in the upper sieves, the seed is somewhere in one of the middle sieves, and the fine rubbish is lower down, or even at the bottom. However, soil sieves do not work well with wet seed, even when used with lots of water. This is because they tend to get clogged rather quickly.

SEED STORAGE

The dried seed should be stored in an air-tight jar which also contains a small bag of freshly dried silica gel. These bags of silica gel are often packed with electronic equipment to keep it dry during transport. They can usually be purchased from a pharmacy, or from a scientific goods supplier. The gel is best dried in a convection oven using the vegetable dehydration cycle. Alternatively, the bag should be left in the hot sun for a few hours. Dryness is essential in seed preservation, and damp seed will not remain viable for long, even when refrigerated. The jar should be labelled to identify both the batch of the seed and its age. If it is stored in a refrigerator at about 4°C, it will keep in a viable condition for several years.

SEED SOWING

When there is no further risk of frost, the seed should be sown in a seedbed with a minimum separation of about two inches between seedlings. Various hand-operated seed drills are available for this sort of work. The seed bed should have a very fine tilth, having been rota-tilled several times in order to germinate and kill weeds. Sow most of the seeds you have been able to produce, keeping back a few as an emergency reserve in case of disaster. It is a good idea to aim at sowing at least 100,000 seeds. In the later breeding cycles, the number of seedlings can be gradually reduced, and this is a point that will be determined by experience.

SEEDLING SCREENING

In the early breeding cycles, there is a grave risk that the blight and beetles will destroy every single plant. The damage will be comparable to the destruction caused by blight, when it first appeared in Ireland. This is obviously a critical stage, and the relatively few survivors must be rescued in good timer. They should be transplanted into pots, and taken to a greenhouse where they can be nursed back to health. It may also be necessary to use crop protection chemicals, both in the field and the greenhouse, in order to save the few survivors. This is a point that should be noted by breeders working on an organic farm, who may prefer to conduct their early breeding cycles on rented land in a conventional farm, where crop protection chemicals can be used without jeopardising their organic certification.

LATER BREEDING CYCLES

In the later breeding cycles, the numbers of seedlings can be reduced. There may be so many surviving seedlings that the best 10-20% would have to be transplanted to a potato field, with a spacing similar to that of a commercial potato crop, before final selections can be made. Professional potato breeders reckon that about 10,000 seedlings must be screened to get a new clone of commercial quality, but this is without regard to horizontal resistance. As the breeding progresses, therefore, and the levels of horizontal resistance increase, the number of seedlings screened can be reduced, but the amount of work required for each seedling will increase. It is at this stage that the breeding begins to get exciting.

INOCULATION AND INFESTATION

In the later breeding cycles, there will be so much resistance that inoculation may be necessary. One or more rows of susceptible modern cultivars should be planted on each side of the seedbeds (or the transplanted potato seedlings) so that blight and beetles can invade the screening populations. It is at this stage that much stronger selection pressures for other selection criteria begin to be applied. These include resistance to other parasites, such as *virus* diseases, as well as attributes of yield and quality. It will probably be advisable to get technical help with the virus diseases.

NUMBER OF GENERATIONS

The method of potato breeding recommended here is based on the behaviour of maize in tropical Africa following the accidental introduction of a rust disease from Central America. When this maize disease first appeared, the loss of crop was almost total, and the farmers had barely enough seed to sow the next crop. This represented the minimum level of horizontal resistance. There are two crops each year in this area and, after 5-7 years, the level of horizontal resistance was maximal, and the loss of crop was negligible. This indicates how many breeding cycles will be required, as well as how very effective transgressive segregation can be.

Note that this maize accumulated all the resistance it needed without any help from professional breeders or pathologists. It could do this because maize is open-pollinated, and the recurrent mass selection was effectively a natural process, assisted by the wisdom of the farmers who were persisted in trying to preserve their favourite landraces. While this accumulation of horizontal resistance was happening, an official maize breeding program was conducted to produce maize with single-gene resistances, but these vertical resistances broke down so quickly that the new lines never even reached the farmers. Breeding potatoes for comparable levels of horizontal resistance to blight, beetle, and other parasites, may take a little longer than the African maize, because more species of parasite are involved. But the difference is unlikely to be large.

ON-SITE SCREENING

On-site screening means that the screening is conducted in the *agro-ecosystem* of future cultivation, in the *time of year* of future cultivation, and according to the *farming system* of future cultivation. This is important because the epidemiological *competence* of parasites varies considerably from one agro-ecosystem to another. A cultivar with horizontal resistances that are in perfect balance with one agro-ecosystem will then have too much resistance to some parasites, and too little to others, when taken to a different agro-ecosystem, or when cultivated out of season, or when cultivated with different cultivation practices.

Fortunately, with potatoes, the agro-ecosystems are usually large, and this is not a serious problem. For example, bacterial wilt (*Pseudomonas solanacearum*) is a tropical disease which lacks epidemiological competence entirely outside the tropics and sub-tropics, apparently because it cannot survive a winter. There is no need to breed for horizontal resistance to this disease in temperate regions. Equally, the temperate virus diseases of potato seem to lack epidemiological competence in the tropics, and there is no need to breed for horizontal resistance to these diseases in tropical countries. However, the principle of on-site screening becomes important with greenhouse work. Greenhouse screening is permissible when there is a 180-day breeding cycle, but only if it is alternated with an on-site field-screening at the proper time and place.

LOCALLY IMPORTANT PARASITES

Because of the need for *on-site screening*, a potato breeding club must be located in the area of future cultivation. The breeding for horizontal resistance can only embrace potato parasites that are *locally* important. If a parasite species is absent from the agro-ecosystem in question, or if it has a very low epidemiological competence, it is not feasible to breed for resistance to it. Nor is there any necessity to do so.

180-DAY BREEDING CYCLE

In the temperate regions, it is normally possible to achieve only one breeding cycle each year, with the cross-pollination and *seed production* undertaken in a greenhouse, during the winter, and the on-site screening conducted in the field, during the summer. With ambitious breeding targets (e.g., horizontal resistance that is complete and comprehensive against all locally important parasites), this could require 10-15 breeding cycles during the same number of years. If, however, we could squeeze two breeding cycles into each year, the total breeding time would be halved. This would necessitate a 180-day breeding cycle, with each cycle divided into a 90-day seed production cycle, and a 90-day screening cycle. This would require a lot of careful planning, and hard work, but it is feasible. It would be achieved by considerable over-lapping of the cycles.

For example, about five times the required number of parents could be selected about two thirds of the way through the screening cycle (i.e., after 60 days of screening). These would be grafted on to tomatoes (sown well in advance) on the clear understanding that 80% of them will be discarded when the final selections are made. By the end of the screening cycle, the selected parents will be grafted, flowering, and ready for cross-pollination. This would add 30 days to the seed production cycle, which would then be effectively 120 days. One breeding cycle would obviously have to be conducted during the winter and, in most temperate countries, this would mean a fairly large heated greenhouse, as well as the induction of both insect infestations and disease epidemics. Technical help may be needed here. The screening cycle would begin by sowing the true seed immediately after it is becomes available at the end of the seed production cycle, including seed taken from somewhat immature fruits.

In the temperate regions, one 180-day breeding cycle should start in spring, and allow an on-site screening in the summer. The second breeding cycle would start in the autumn, and this would necessitate a winter screening in the greenhouse. Obviously, this winter screening would not be an on-site screening but, as this happened in every alternate breeding cycle, the small reduction in breeding efficiency would be acceptable, in view of the huge gain in time.

SOURCES OF ERROR

A number of common errors can occur in the understanding of horizontal resistance, and they must be clearly recognised by the amateur breeder, because they can be dangerously misleading. The first concerns the belief, held by many organic farmers, that organic farming produces plants that have an improved resistance to many parasites. However, good nutrition affects the *physiology* of the plant, but it does not affect the *genetics* of the plant. And resistance to parasites is controlled genetically. Anyone who doubts this should try growing a modern cultivar of potatoes in, say, North America or continental Europe, using the best possible soils and nutrition, but without spraying against either *blight* (*Phytophthora infestans*) or *Colorado beetle* (*Leptinotarsa decemlineata*).

There is an important difference between a low incidence of a parasite, and resistance to that parasite. A low incidence means that the parasite is rare in the crop in question, but this could change at any time, and the parasite would then become common. A high resistance means that the parasite is rare, and it remains rare, in that crop. Many organic farmers can grow unsprayed potatoes reasonably successfully because of a low *incidence* of blight and beetle. This low incidence results either from their isolation, or from the fact their neighbour's potato crops are all being sprayed. And it is then very easy, and very tempting, to attribute the reduced parasitism to an increase in resistance resulting from a healthy soil and organic farming. If potatoes are grown organically in large quantities, the incidence of both blight and beetle will rise sharply, and the need for increased levels of horizontal resistance to these parasites will become critical.

A second error concerns chance escape from parasitism, and this happens most with slow-moving parasites such as virus diseases. These individuals are susceptible but, being parasite-free, they look so healthy that they are mistakenly believed to be resistant. If they are used as parents, there will be little breeding progress in accumulating horizontal resistance. There are two methods of avoiding this error. One is to select only plants that have slight parasitism, rather than no parasitism. The other is to *inoculate* the screening population.

A third error is to confuse *vertical resistance* for horizontal resistance. Fortunately, with potatoes, this is not normally a serious , and because the other parasites, in which vertical resistance occurs, are rare. If a functioning vertical resistance to blight is suspected, look for the small necrotic flecks, just visible to the naked eye, which indicate that the hypersensitivity mechanism of vertical resistance is operating. problem with *blight,* because of the presence of the *A2 mating type* and because the other parasites, in which vertical resistance occurs, are rare. If a functioning vertical resistance to blight is suspected, look for the small necrotic flecks, just visible to the naked eye, which indicate that the hypersensitivity mechanism of vertical resistance is operating.

A fourth error concerns biological anarchy, which means that the natural biological controls have been severely damaged, even destroyed, by the use of crop protection chemicals. Because potato seedlings must be screened without any crop protection chemicals, the parasites will behave with a savagery that would be impossible if their biological controls were functioning fully. This is another reason for using relative measurements of horizontal resistance. It is also a clear indication that we will need less horizontal resistance than we may think, once the biological controls have been restored. The best way to restore the biological controls is to use horizontal resistance. And the best way to enhance the effects of horizontal resistance is to restore the biological controls. The two effects are mutually reinforcing.

A fifth error concerns parasite interference, which results from the fact that most parasites are mobile, and can spread from one plant to another. If a resistant plant in a screening population is surrounded by susceptible plants, parasites will move on to the resistant plant and make it appear much more susceptible than it really is. This is another reason for using relative measurements of horizontal resistance. This error also suggests that we will need much less horizontal resistance than we may think, when our seedling becomes a clone grown in a farmer's field without any parasite interference at all.

Finally, there is the error caused by 'population immunity'. This error concerns *population* growth rates which can be either positive or negative. If a population exhibits more births than deaths, the population growth rate is positive, and the population increases in size. Conversely, if the deaths exceed the births, the population growth rate is negative, and the population decreases in size. If a parasite is to cause an epidemic, its population growth rate must obviously be strongly positive. However, there may be a level of horizontal resistance such that the parasite population growth rate is close to zero, or even negative. This situation is called population immunity, because the epidemic does not develop, even though the host individuals within that population are less than immune.

The error arises when host individuals are believed to be susceptible, because the population immunity is not apparent under the conditions of measurement (e.g., in a greenhouse or plant growth chamber). Once again, we may need less horizontal resistance than we think, because population immunity can prevent an epidemic from developing in a cultivar that is less than completely resistant.

MEASURING HORIZONTAL RESISTANCE

The level of horizontal resistance can be determined only from the level of parasitism. Various factors, such as the sources *of error* just described, can increase the levels of parasitism very considerably in a screening population, or in research plots. For this reason, the level of horizontal resistance is likely to *appear* much less than it really is. But this appearance is deceptive. Indeed, this deception is probably the reason why professional plant breeders have ignored horizontal resistance so consistently. It is very unlikely that we shall ever have a scale of measurement for horizontal resistance comparable, say, to the Celsius scale for temperature. In other words, we cannot make *absolute* measurements of horizontal resistance. We can make only *relative* measurements. These relative measurements are important in two situations.

First, when screening potato seedlings for horizontal resistance, it is important to select the *least* parasitised individuals, regardless of how severe their parasitism may be. When grown as a clone in farmers' fields, without parasite interference, and with restored biological controls, these apparently susceptible seedlings may well have population immunity. At the very least, they would make the best parents for the next breeding cycle. It would be tragic if such seedlings were destroyed, just because they *appeared* to be so much more susceptible than they really are. Second, a new cultivar has to be described, and it is important that its horizontal resistances to major parasites should be accurately indicated. The only way to do this is to compare it with well-known commercial cultivars. That is, for each important pest or disease, you can say that its horizontal resistance is superior to cultivar A, and inferior to cultivar B. An alternative approach is to indicate how much spraying, if any, and with what, is needed in an average season.

Horizontal resistance can also be determined with crop loss assessment trials, but these trials are complicated and they are not recommended for amateur breeders. Such determinations enable you to say what sort of yield loss each parasite might cause in an unsprayed crop, in an average season.

TUBER RESISTANCE

Blight, as well as a number of soil-borne pests and diseases can attack the tubers. Tubers of the more promising parents, particularly in the later breeding cycles, should be kept and cultivated as small clones. Tuber resistance to these various parasites can then be assessed fairly easily. Clearly, tubers that rot quickly in store are not very resistant, and they are not very useful, either. But, here again, all measurements of resistance must be relative measurements. This need for measuring tuber resistance will require a longer breeding cycle and, at this stage, it may be as well to revert to the one-year breeding cycle.

TUBER QUALITY AND YIELD

Visible characteristics of tuber quality include tuber shape and size, and the depth of eyes, deep eyes being difficult to peel. The colour of the skin and flesh can vary considerably, and local preferences can have a big effect on sales. For example, people in England generally prefer white flesh potatoes, while people in Holland prefer yellow flesh. Tuber yield is related to the number of tubers, and the size of tubers. Generally speaking, fewer large tubers are preferable to the same weight of smaller tubers. Manufacturers of French fries like large, long tubers. Tubers are sorted by size with a 'riddle', which is a coarse sieve, and the smaller tubers can be kept for 'seed'.

COOKING CHARACTERISTICS

Cooking tests can be made only after a sizeable quantity of tubers from each promising seedling has been accumulated by vegetative propagation. These tests include taste, tuber colour (white or yellow), texture (floury or soapy), and specialised uses, such as salad potatoes, which must not turn black when cold, French fries, roasting, baking, boiling, and starch production. Most of these characteristics are quantitatively variable. The demands of commercial markets can be important, as with French fries in fast food outlets, and the manufacture of potato crisps. Many industrial countries have special laboratories that can make an official report on these matters, including dry weight, starch content, protein content, vitamin C, glyco-alkaloids, etc. Depending on the country, this reporting can be expensive, but it is very useful during the process of registration.

AGRONOMIC SUITABILITY

The two principle characteristics of agronomic suitability are stolon length and leaf spread. Short stolons are necessary for mechanical harvesting. They are less necessary for subsistence cultivars but, even so, short stolons make hand harvesting easier. A good spread of dense leaves is a valuable aid to weed suppression and, for this reason, potatoes are often called a cleaning crop. However, such a leaf spread will normally develop only from a seed tuber, and it cannot be expected from a true seedling. A mass of leaves is also a useful screening character, because a mass of *green* foliage is a very good indication of resistance. While a mass of foliage does not necessarily indicate a high yield of tubers, it is usually a prerequisite for such a high yield.

FLOWERING IN POTATO CROPS

Many potato cultivars produce flowers and fruits in cultivation. These fruits constitute a 'physiological sink' which take a large share of the nutrients being produced by the 'physiological source' of the leaves. Cultivars that flower early and profusely, and set many fruits, will yield fewer and smaller tubers. Consequently, the character of 'late flowering', or 'no fruits', or even 'no flowering' is a useful, selection criterion, even though it might lead to difficulties in breeding.

RAPID MULTIPLICATION

A promising new selection should be multiplied as rapidly as possible in order to provide material for various tests, and registration. A useful rapid multiplication technique involves green stem cuttings. A piece of potato stem is used as a cutting. It is put in a mist propagator for about five days until roots develop . It is then planted in soil, in a pot, still in a mist propagator, until vigorous growth commences. In the meanwhile, the stem from which the cutting was taken will produce two new side-shoots, which can also be used as cuttings. These parent stems will then develop four new shoots, and so on until the parent plant looks like a candelabra . This technique has even been called the 'candelabra technique'. The rooted cuttings will themselves produce cuttings, and begin to resemble candelabra. All these plants will also produce tubers that can be used for rapid multiplication, in both the greenhouse and the field.

MIST PROPAGATION

Cuttings are most likely to form roots if: (i) they have plenty of leaf; (ii) they are exposed to strong light; (iii) they have a rooting medium that is biologically and nutritionally inert; and (iv) they are kept in a 100% humidity. These conditions are met in a simple apparatus called a mist propagator. When the cuttings have plenty of leaf, and plenty of light, they can manufacture all the carbohydrates they need in order to grow roots. But they must be kept moist until they are able to produce enough root tissue to absorb water from the rooting medium. The rooting medium must be biologically and nutritionally inert to prevent rot-causing fungi and bacteria from attacking the cuttings. The mist propagator consists of a tray containing the rooting medium. A good rooting medium is polystyrene granules mixed with sand in a ratio of about 4:1. If polystyrene granules are not available, use plain sand, or sand mixed with vermiculite or perlite. The tray is enclosed in a tent of transparent plastic film, and it may be exposed to full sunlight, at least for part of the day. A domestic humidifier is used to continuously pump a fine water mist into the tent. Alternatively, a hand sprayer can be used every hour or so during the day, in order to keep the inside of the mist propagator permanently moist. A more simple apparatus consists of a plastic bag tied over the top of a flowerpot and supported by sticks. Under mist propagator conditions, potato cuttings will form roots in 5-10 days. They can then be transplanted into pots of soil, and protected by a plastic bag cover for a few days. Once they are growing vigorously, they can be transplanted into larger pots or into the field.

DAY-LENGTH

Potatoes originated in the high altitude tropics of South America, where the days and nights last for about twelve hours throughout the year. Consequently, they are 'short-day' plants. This means that various growth processes such as flowering and tuber initiation will occur only under the influence of a twelve-hour day. Such potatoes cannot be cultivated in the temperate regions because tuber initiation does not occur until the autumn equinox in late September (or late March in the southern hemisphere), and the plants are killed by winter before they are ready to harvest. About two centuries of both natural and artificial selection in Europe produced day-neutral plants that are able to initiate tuber formation in the long days of a temperate summer. These are day-neutral plants rather than long-day plants. They can be successfully cultivated in the tropics, but they are likely to very susceptible to the tropical disease called bacterial wilt (*Pseudomonas solanacearum*). It follows that amateur breeders in temperate countries should use only day-neutral plants in their breeding program. But amateur breeders in the tropics may use potatoes with any degree of photoperiod sensitivity.

GLYCO-ALKALOIDS

Potato leaves and fruits are poisonous, because they contain toxins known as glyco-alkaloids (i.e., leptinines, leptines, solanine, and chaconine). Sometimes, the tubers contain sufficient concentrations of glyco-alkaloids to prevent their registration as new cultivars. Without worrying about this too much, potato breeders should keep an eye on glyco-alkaloid concentrations, and clones with high levels in the tubers should not be used as parents. Testing for glyco-alkaloids is not difficult, but it must be done in a professional laboratory, such as a university or a commercial company, usually for a fee.

WIDENING THE GENETIC BASE

In the later breeding cycles, the rates of gain in quantitative variables, including horizontal resistances, may decline before the breeding targets have been achieved. This would happen because the genetic base of the original parents was too narrow. Such an error is easily corrected by adding new genetic material to the breeding program. This is likely to lead to a small loss in various quantitative variables, but these losses will soon be made up, and the breeding ceiling will be raised considerably. The use of neo-*tuberosum* material is recommended for such widening. However, it must be clearly recognised that there is a limit to every quantitative variable and, once that limit is reached, no further progress is possible. Most of these limits are unknown because so little quantitative breeding has been undertaken in potatoes. But the amateur breeder may safely assume that these limits represent levels of horizontal resistance that are vastly superior to those of modern cultivars. The limits to other variables, such as yield and quality, are also unknown, but modern cultivars are obviously much closer to them than they are in their horizontal resistances.

OLD ENCOUNTER, NEW ENCOUNTER, AND RE-ENCOUNTER PARASITES

Old encounter parasites evolved with their host, and they have been in contact with it since its original domestication. This happens in the centre of origin as well as in new areas, when both the host and the parasite were taken there together.

New encounter parasites evolved on a botanical relative, but they remained out of epidemiological contact with their domesticated host. They were brought into contact with their domesticated host by people.

Re-encounter parasites evolved with their host, but they were separated from it when people took the host to another part of the world, leaving the parasite behind. At a later date, the parasite is inadvertently introduced to the new area where it re-encounters its host.

Blight, Colorado beetle, and the temperate *virus* diseases are new encounter parasites of potato. This may mean that a few more breeding cycles will be necessary than if these were old encounter parasites. However, it would be quite wrong to conclude that it is impossible to obtain adequate levels of horizontal resistance to new encounter parasites. High levels of horizontal resistance to blight have been clearly demonstrated. The fact that no one has achieved high levels of horizontal resistance to Colorado beetle is because such breeding is only now being attempted. And it seems that no work has been attempted with horizontal resistance to the temperate virus diseases of potato. Most other potato parasites are re-encounter parasites. The host tends to lose resistance in the absence of a parasite. If the separation between host and parasite has been lengthy, the loss of horizontal resistance may be considerable. The classic example of this occurred with the maize of tropical Africa.

BLIGHT

Potato blight, caused by the microscopic fungus *Phytophthora infestans*, is by far the most important disease of potatoes, and it now occurs in every potato growing area of the world. In about 1840, the potatoes on a ship travelling from Mexico to New York mysteriously rotted. This ship was inadvertently carrying the blight fungus from its centre of origin in Mexico to the temperate Northern Hemisphere. Shortly after this, the cook on another ship, travelling from New York to Rotterdam, also found that all his potatoes were rotten. This ship carried the blight from the New World to Europe. The fungus then spread rapidly and, by 1845, it had reached Ireland. The damage it caused was incredible, and potato crops all over Europe were reduced to a black, stinking mush. This was the first devastating plant disease known to history, and it was devastating because it was a 'new-encounter' disease. The potato had evolved and been domesticated in South America, while the fungus had evolved on botanical relatives in Mexico, and they had been brought together by people. The potato had very little resistance to the fungus.

In Ireland alone, about one million people died of starvation, and another one and a half million emigrated, mainly to North America. This loss of people reduced the population of Ireland by one third. It is thought that another million people died in the rest of Europe. The decade became known as "The Hungry Forties". Then, after two or three years, the blight epidemics declined. For the next forty years, potatoes were cultivated all over Europe and North America. Although blight limited the yields, potatoes remained one of the most important food crops in the Northern Hemisphere. The reason for this decline in the epidemics was the rapid and complete elimination of all the most susceptible clones of potato. Only relatively resistant clones remained. Furthermore, all new potato breeding automatically produced resistant clones, because every susceptible seedling was killed by blight. This was horizontal resistance, and it enabled massive potato cultivation to continue without any spraying with fungicides. Indeed, no fungicides were known at that time. Equally, no vertical resistances were known at that time.

In 1882, Millardet, in France, discovered the fungicide that he called *bouillie bordellaise*, or Bordeaux mixture, which consisted of a mixture of copper sulphate solution and newly slaked lime. He used it to control downy mildew of grapes, which had also been accidentally imported from the New World, and was threatening the wine industry with ruin. And it was quickly discovered that this fungicide would also control potato blight. This made potato cultivation easier and more profitable. But it ruined potato breeding. This was because the breeders could now spray their screening populations, and this made the breeding very much easier. All the most famous potato varieties were produced during this period which lasted another forty years. Unfortunately, during this breeding, the accumulation of blight resistance not only stopped. It went into reverse, and horizontal resistance was gradually lost because of the *'vertifolia effect'*. This loss of resistance was not recognised at the time and, even if it had been, it would have been considered unimportant, because the crops could be sprayed with Bordeaux mixture. The slow loss of quantitative resistance was not even noticed, and potato varieties bred after about 1885 became increasingly susceptible.

During World War I, Germany was critically short of many commodities but, in particular, copper was scarce. Copper was needed for making Bordeaux mixture. But it was also needed for making brass shell cases for the rifles and field guns. Because of the war, the military had priority, and the potato crops went unsprayed. The winter of 1917 was known as the 'turnip winter'. Germany lost the war mainly because of food shortages, and several countries decided that potato blight had military significance. When the war was over, they began to breed potatoes for resistance to blight. And they used the most modern techniques available. They used single-gene resistances that obeyed the recently recognised Mendel's laws of inheritance. In a word, they used vertical resistance.

For the next forty years, there was great optimism, as scientists believed they would eliminate blight forever. But these *unstable* vertical resistances failed again and again. And, during the 1960s, the breeders decided that it was a waste of time breeding for vertical resistance to blight. They should have tried horizontal resistance but they did not, probably because they were so badly misled by various sources of error. A notable exception to this rule was John S. Niederhauser, who was working in Mexico. He was the first scientist who deliberately avoided the use of vertical resistance and worked with horizontal resistance. He was very successful and, in 1991, he was awarded the World Food Prize, which is the agricultural equivalent of the Nobel Prize.

A2 MATING TYPE OF BLIGHT

For about a century, scientists were baffled by the fact that the blight fungus apparently had no sexual reproduction. The problem was solved by Jorge Galindo, a famous Mexican scientist. Galindo showed that the blight fungus has two mating types. Each mating type is hermaphrodite (i.e., it has both sexes) but it is self-sterile. This means that both mating types must be present if the blight fungus is to reproduce sexually. Sexual reproduction has two advantages for the fungus. First, it produces great variation in the fungal population and, second, it produces very tough 'oospores' that can survive outside the host during a winter, a drought, or an otherwise rough environment, without being killed. If only one mating type is present, the fungus can only reproduce vegetatively, by means of very fragile, asexual spores. It can then produce variants by mutation only, and it can survive a winter only inside its host. That is, it can survive only in potato tubers.

When the blight fungus was taken from Mexico to New York, and then to Europe, it was taken as one mating type only, now known as A1. For about 150 years the entire population of the blight fungus in the temperate regions consisted of this one mating type only. But, recently, the second mating type, discovered by Jorge Galindo, and now known as A2, was accidentally taken from Mexico to Europe. Before its presence was detected, it had been spread all over the northern hemisphere in certified seed potatoes. The presence of A2 means that the blight fungus can now form oospores in huge numbers. This has several effects on the blight epidemics.

First, the greatly increased variability in the fungus means that vertical resistances will break down even more quickly, and that they are even less useful than before. And modern synthetic fungicides, such as glyphosate, will also break down more quickly.

Second, when only A1 was present, the blight could survive a winter only in potato tubers. Blighted tubers were relatively rare, and this meant that the initial inoculum was small, and the epidemics required more time to develop fully. The disease was known as 'late blight' for this reason. But, with huge numbers of oospores in the soil, the initial inoculum becomes very large, and the epidemics start earlier, and they are more serious. This means that higher levels of horizontal resistance will now be necessary if the blight is to be controlled by breeding.

Third, when only A1 was present, tomatoes could get blight only from diseased potatoes, and this made it 'late blight' on tomatoes also. But tomatoes can now get blight directly from oospores, and they get it much earlier. The disease is now much more severe on tomatoes, and some plant breeding clubs might care to breed tomatoes for comprehensive horizontal resistance, in parallel with their potato breeding.

Fourth, areas that are still free of the A2 mating type should be very careful not to import it. It is certain to reach all such areas eventually, but any delay will be useful, until such time as local cultivars with adequate resistance to blight have been produced.

There are two advantages obtained from the presence of the second mating type. First, a dangerous crop vulnerability has been realised and, once overcome, it will be permanently eliminated. Second, the inactivation of vertical resistances to blight is so rapid that the one-pathotype technique is no longer necessary during potato breeding.

COLORADO BEETLE

As settlers moved west, in the U.S.A., during the second half of the nineteenth century, they eventually reached the State of Colorado. A harmless beetle that parasitised the wild *Solanum rostratum*, known as prickly potato, or the buffalo burr, moved into their potato crops and it became one of the worst crop parasites ever known. Like blight, it is a 'new encounter' parasite. Its Latin name is *Leptinotarsa decemlineata*, and it was given the English name of Colorado beetle, after its home state, and its centre of origin. There are many areas, such as the United Kingdom, Africa, Central and South America, and the Pacific Northwest of North America, where the Colorado beetle does not occur. Anyone breeding potatoes in such areas obviously cannot accumulate horizontal resistance to this insect pest. And any new potato cultivars produced in such an area will have a commercial value limited to those beetle-free areas. Conversely, beetle-resistant potato cultivars will be useful in beetle-free areas, even if their resistance is unnecessary. These limitations should be borne in mind by anyone thinking of starting a potato-breeding club.

David Fisher is breeding potatoes with success, in the U.S.A., for horizontal resistance to Colorado beetle, and one of his colleagues has shown that the horizontal resistance is not due to a high glyco-alkaloid concentration. He also makes the useful observation that, because potatoes can lose up to one third of their leaf tissue without significant yield loss, it may be economically feasible to control the losses caused by Colorado beetle with levels of horizontal resistance that are far from maximal. Some insect larvae, such as the Monarch butterfly, which feeds on poisonous milkweed, and the Colorado beetle, which feeds on poisonous potato leaves, are able to isolate these poisons in a special sac inside their bodies. This makes both the larvae and the adults poisonous to insect-eating birds. From their colouring, insect-eating birds recognise these poisonous insects and avoid them. Consequently, putting chickens in the potato patch will not provide a biological control of Colorado beetle.

VIRUS DISEASES

The scientific literature reveals that little potato breeding has been undertaken for resistance to virus diseases. The reason is fairly simple. Single-gene resistances to these diseases are relatively rare, and even when they exist, they provide unstable resistance. It was also argued that the use of certified seed potatoes made their development unnecessary. There was also a fear that their development would damage the certified seed industry. Additionally, potato breeders were not working with quantitative resistance. Virus diseases spread relatively slowly in a screening population, and any plant that got a virus diseases would be discarded, on the grounds that it was susceptible. The potato seedlings that were kept were not resistant, however. They were all susceptible 'escapes'.

Consequently, there is immense scope for amateur potato breeders to produce cultivars that are largely unaffected by these viruses, or any other tuber-borne diseases. Farmers can then keep some of their own harvest for seed. This would reduce the costs of potato production enormously. The certified seed potato industry would then be reduced to providing elite seed stocks of new cultivars, certified as to identity and purity of variety only. Anyone who doubts that this is possible should visit Kenya, where some potato cultivars have been grown for more than sixty vegetative generations without any renewal of seed stocks. The major seed tuber-borne diseases are:

1. *Potato Leaf Roll Virus.* Insect transmitted by *Myzus persicae* and other aphids. The leaves roll into cup shapes, even cylinders, and become brittle.
2. *Potato Virus X.* Mechanical transmission.
3. *Potato Virus Y.* Also known as 'Leaf Drop Streak'. Main spread in field is by *Myzus persicae* but also sap-transmissible.
4. *Spindle Tuber Viroid.* Transmitted in true seed.
5. *Other Potato Viruses.* Generally unimportant.

OTHER TUBER-BORNE DISEASES.

Readers should consult *The Compendium of Potato Diseases.* Screening for horizontal resistance to tuber-borne diseases is done by visual symptoms on the tubers, and by tuber-rotting in store. These diseases include:

1. Ring rot (*Corynebacterium sepidonicum*).
2. Black Leg (*Bacterium carotovorum*).
3. Common Scab (*Actinomyces scabies*).
4. Powdery Scab (*Spongospora subterranea*).
5. Wart Disease (*Synchytrium endobioticum*).
6. Golden nematode (*Globodera rostochiensis*).
7. Bacterial wilt (*Pseudomonas solanacearum*).

EPIDEMIOLOGICAL COMPETENCE

Epidemiological competence is the ability of a parasite to produce an epidemic. The epidemiological competence is usually quantitative, in the sense that it can vary between a minimum and a maximum. This occurs with bacterial wilt (*Pseudomonas solanacearum*) of potatoes, for example. The epidemiological competence of this bacterium is maximal in the tropics, and it declines steadily towards the subtropics. In a region with frost in the winters, the epidemiological competence is lost entirely. Equally, the temperate virus diseases seem to lack epidemiological competence in the tropics.

Variation in epidemiological competence is the reason for *on-site screening*. New cultivars must have horizontal resistances that are compatible with the epidemiological competence of all the local parasites in the agro-ecosystem of future cultivation. If the new cultivar was selected in a different agro-ecosystem, it will have too much resistance to some parasites and too little to others, when brought to its new agro-ecosystem. Having too much resistance does not matter, but having too little resistance represents a failure of the breeding program. Fortunately, most potato agro-ecosystems are large, and a new potato cultivar is likely to be agriculturally suitable over a considerable area.

ONE-PATHOTYPE TECHNIQUE

This is a technique for ensuring that vertical resistances are inactivated during screening for horizontal resistance. Amateur breeders are unlikely to need this technique because the A2 mating *type* makes it unnecessary with *blight*, and other diseases with vertical resistances (e.g., wart disease, golden nematode) are unlikely to feature in amateur potato breeding programs. Should they need it, a full description of this technique is available in *Return to Resistance.*

ILLEGAL PARASITES

Some potato parasites are the subject of legislation and it may be illegal to work with them. These include areas where Colorado beetle is still absent, such as the United Kingdom, Mexico, and the Pacific Northwest in North America. Wart disease (*Synchytrium endobioticum*) and golden nematode (*Globodera rostochiensis*) are also controlled by legislation is many countries.

CROP VULNERABILITY

Crop vulnerability means that a crop is susceptible to a foreign parasite which, however, is absent from the region in question. When that parasite is accidentally introduced, the vulnerability is manifested, and potential parasitism becomes actual parasitism. Crop vulnerabilities can vary from the trivial to the catastrophic. Historically, two of the worst crop vulnerabilities were the potatoes of Europe prior to the introduction of blight and Colorado *beetle*. Two more were the European wine grapes susceptible to *Phylloxera* and downy mildew, which were both new-encounter parasites accidentally imported from the New World.

Crop vulnerability is important to amateur plant breeders in two ways. First, they should make themselves aware of the crop vulnerabilities in their region, and they should be very careful not to introduce foreign parasites on imported breeding stocks. They should familiarise themselves with their country's phytosanitary regulations and take great care to obey them. This is particularly true of breeding clubs in island nations, such as the United Kingdom, Australia, Madagascar, etc. Second, a serious crop vulnerability may be realised during the breeding activities of a club, and their new cultivars would then be susceptible to the foreign parasite. This can be very disappointing as promising new cultivars would be ruined. However, the club members should immediately make horizontal resistance to the new parasite one of their selection criteria, provided that this was *legal*. They may well be the first to solve this new problem, and to provide a valuable service to their region.

CULTIVAR FORMATION

The biological formation of a new cultivar of potato is easy because this crop is cultivated as a clone, propagated by tubers. However, the tubers emerging from a breeding program will be either infected or contaminated with pests and pathogens, including virus diseases. In the industrial countries, potato seed legislation will usually demand propagating material shown to be free of many specified pests and pathogens. In practice, this usually necessitates a technique called 'meristem culture', in which new plants are produced from parasite-free 'stem' cells. This is not a technique for amateur breeders, and it would normally be undertaken by the registering authority. These days, cultivar identification is usually done by DNA-fingerprinting, and this process also is undertaken by the registering authority.

CULTIVAR REGISTRATION

The legal requirements of cultivar formation concern registration of the cultivar, licensing of seed growers, collection and payment of royalties, etc. These requirements vary from country to country, and you should familiarise yourself with your own country's legislation. Many tropical countries have no legislation of this kind and the amateur breeder must give his new cultivars to farmers for trial and approval. Successful cultivars will spread rapidly.

ROYALTIES

The payment of breeders' royalties is based on the sale of certified seed tubers of a registered cultivar. Both the duration of the copyright, and the amount of royalties vary from country to country. But even a moderately successful potato cultivar will be grown on many acres, and could earn many thousands of dollars a year in royalties. In the non-industrial world it is often impossible to register a cultivar, or to collect royalties, because there are no certification schemes, and no mechanisms for collecting royalties. And a breeding club should not want to earn royalties from subsistence farmers. As a general rule, breeding clubs in the Third World cannot expect to earn royalties, and their efforts should be entirely altruistic. Potato breeding clubs should work for both the elimination of certified seed, and the elimination of crop protection chemicals. Ironically, the more successful they are in these targets, the more they will reduce their potential royalties. However, potato breeding clubs are intended to function for the betterment of humankind, rather than for the gaining of filthy lucre.

FARMER'S PRIVILEGE

In the plant breeders' rights legislation of most countries, there is a clause called 'farmer's privilege'. This means that a farmer is entitled to keep some of his own harvest of a copyrighted cultivar for seed, but only for his own use. He may not sell any of this seed unless licensed to do so.

BREEDER'S PRIVILEGE

In most countries, plant breeders are allowed to use copyright cultivars as parent material. If in doubt, check with your agricultural authorities. In some countries, where single genes are still considered important, it is possible to patent a single gene. This is of no interest to amateur breeders who will normally be working exclusively with quantitative variation (i.e., polygenes).

SEED CERTIFICATION

There are two kinds of potato seed tuber certification. The traditional certification is undertaken by government inspectors who confirm that crops grown for seed, on licensed seed growers' farms, are free from virus diseases. They also check the trueness and purity of variety. The second kind of certification concerns freedom from synthetic chemicals in seed tubers for organic farmers. In many countries it is becoming possible to have a double certification covering both aspects of the guarantee.

ELIMINATION OF CERTIFIED SEED TUBERS

The cost of certified seed tubers is high, and is probably the largest single input in potato cultivation. If amateur breeders produce new cultivars so resistant to tuber-borne parasites that they do not need certified seed tubers as a means of parasite control, farmers need buy certified seed only if they want to introduce a new potato cultivar to their farm. This will bring down the cost of potato cultivation dramatically, and this should be one of the main aims of plant breeding clubs.

Obviously, this use of resistance will largely eliminate the certified seed industry. However, given proper potato breeding, this industry should never have been necessary in the first place. Nor should the very high cost of potato cultivation have been necessary during the past century or longer. In any event, the certified seed industry is being given plenty of notice, because it will inevitably be some time before such resistant cultivars become available.

A second consideration, already mentioned, is that this elimination of certified seed as a means of parasite control will eventually reduce the breeders' royalties to almost nothing. These royalties are calculated on the basis of sales of certified seed. However, this should be regarded as proof of success, because most plant-breeding clubs will be working primarily for the benefits of food that is both cheap and free from crop protection chemicals. Royalties should be a secondary consideration.

ELIMINATION OF CROP PROTECTION CHEMICALS

The cost of spraying potato crops with crop protection chemicals often equals or even exceeds the cost of using new certified seed for each crop. The frequency of spraying potato crops, and the complexity of the chemical mixtures be applied, have both been increasing during the past decades. This is partly due to changes in the parasites, such as the development of increasing pesticide-resistance, or the A2 mating type of blight. It is also due to the susceptible nature of the potatoes themselves.

Breeding for horizontal resistance is cumulative and progressive. This means that there will be many quantitative improvements in resistance, and the amount of spraying will gradually be reduced as better and better cultivars become available. If the amateur potato breeding clubs are successful, their new cultivars will eventually be so resistant that they will need very little spraying or, possibly, none at all. This will make a dent in the markets of the big chemical corporations. However, given good potato breeding, this spraying should never have been necessary in the first place. And the chemical corporations are being given plenty of notice.

Nevertheless, we should expect some quite fierce opposition from the chemical manufacturers. It will probably be covert opposition, and it is most likely to appear in the form of refusing research grants to universities that work with horizontal resistance. This is one of the reasons that amateur breeding clubs are so necessary.

REFERENCES

Fisher, D.G., Deahl, K.L., & Rainforth, M.V. (2002); Horizontal Resistance in *Solanum tuberosum* to Colorado Beetle (*Leptinotarsa decemlineata* Say). *Amer. J. of Potato Res.* 79: 281-293.

Information: Safety Assessment of Potatoes Resistant to the Colorado Potato Beetle. Health Canada. April 1997.

"Managing Colorado potato beetle not simple". *Home & Garden.* 16 May 1999.http://www.messenger-inquirer.com/homegarden/69092.htm

Simmonds, N.W. (1991); Genetics of Horizontal Resistance to Diseases of Crops. *Biol. Rev.*, 66: 189-241.

10

Organic Breeding of Brambles

The classification *bramble fruits* applies to a large number of plant species within the genus *Rubus*. The most commonly known bramble fruits are blackberries and raspberries, but other types—such as dewberries, tayberries, boysenberries, and loganberries—are also popular. Many of the bramble types and varieties are thorned, though some are not. Plant breeders have worked especially hard in recent years to develop thornless varieties. Growth habit also varies among bramble fruits. Some have prostrate growth and require trellising for commercial production; others grow fully upright and don't need trellising. Typically, bramble canes are vegetative during their first year of growth, and in this stage they are called *primocanes*. They are called *floricanes* the following season, during which they bear fruit. New primocanes emerge while the older floricanes fruit, so mature plants have both active primocanes and floricanes in evidence each season. Several raspberry cultivars are primocane-fruiting. These varieties produce a late summer or fall crop on the upper portions of the primocane stem in the first year. The next year, the same cane—now a floricane—produces a second crop lower on the stem.

Because growth and fruiting habits vary among bramble species, so do production practices. It is important that the grower seek out the cultural guidelines for the specific berry being grown. Most bramble plantings are established either from root cuttings or plants. Spacing within rows at planting depends largely on the species. Eventually, most bramble fruits are managed as continuous hedgerows rather than as individual plants. The row spacing for commercial hedgerow plantings is typically 12 feet to accommodate tractor operations, allow for air movement, and optimise sunlight penetration. Brambles, especially blackberries, are fairly drought-tolerant and are sometimes grown without supplemental irrigation. Most commercial plantings, however, are irrigated to assure quantity and quality of production. Because bramble fruit is highly perishable, marketing strategies must be developed well in advance. Many growers rely on direct marketing—particularly U-pick, roadside stands, and farmers' markets—to sell their crop.

SITE SELECTION AND PREPARATION

A planting site higher than the surrounding area provides air circulation and water drainage, reducing disease and insect problems. Sites with adequate air circulation also reduce the possibility of winter injury and late spring frosts. However, where strong prevailing winds occur, open sites (such as the crests of hills) may expose plants to desiccation and excessive wind stress—major causes of winter kill in brambles. Brambles are adapted to a wide range of soils but do best when planted on well-drained, deep, fertile soil that is high in humus and free of hardpan. A pH of 6.0 to 6.5 is preferred. Compared to raspberries, blackberries are more tolerant of heavy soils and the soil-borne diseases often associated with heavy soils. Adding organic matter and planting in raised beds can be of help in avoiding these problems.

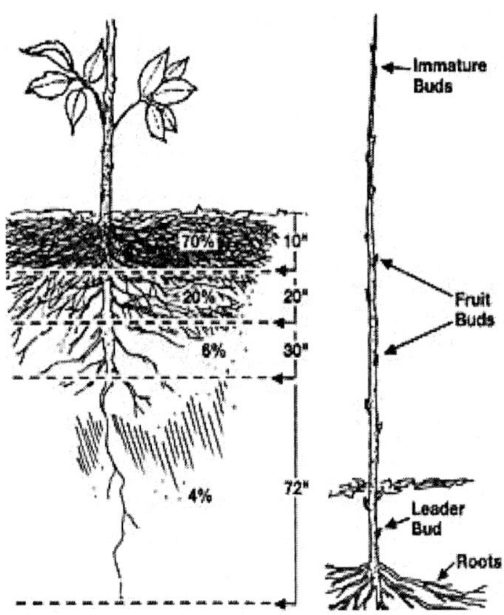

Figure 1. Planting of a Bramble plant

Sites previously established in fruit crops (e.g., peaches, apples, grapes, brambles, etc.) are poor choices because of potential problems with crown gall infection. Likewise, land recently planted to solanaceous crops (e.g., tomatoes, potatoes, peppers, eggplant, tobacco) presents increased risks of verticillium wilt. Nearby thickets of wild brambles should also be removed, if possible, to reduce sources of insect pests and diseases. Experts advise a minimum of 300 to 400 feet of separation from wild hosts or other commercial plantings. Perennial weeds, especially aggressive grasses such as bermuda, should be controlled in advance of planting, or an alternative site should be chosen. This is extremely important in organic and reduced-herbicide plantings, as perennial weeds are especially difficult to manage with non-chemical methods.

SPECIES AND CULTIVAR SELECTION

It is advantageous to organic and low-input growers to select well-adapted species and varieties, especially those genetically resistant to common diseases. By so doing, growers find they have more time and resources to spend on other critical areas, such as weed control. Cultivar information is readily available from state or county Cooperative Extension services and from local nurseries. When purchasing, it is also important to buy from reputable suppliers to ensure virus-free and nematode-free planting stock. With one exception, there are few restrictions on proximity when planting different varieties and species of brambles on the same farm (assuming all are locally adapted). Black and purple raspberries are much more susceptible to damage from mosaic and leaf curl viruses than are red and yellow cultivars. Since these diseases are vectored by aphids, black and purple varieties should be separated as much as possible, and located upwind, from red and yellow raspberries.

Figure 2.An young bramble plant

ALLEYWAY MANAGEMENT

Brambles can be managed either under clean cultivation or with sodded alleyways. In some instances, the alleyways between new plantings are clean-cultivated, with a permanent cover crop planted at a later date. Brambles have extensive root systems, and as plantings mature, deep tillage between berry rows must be avoided to prevent root damage, unwanted suckering, and increased susceptibility to crown gall through plant wounds. According to Kansas recommendations, cultivation should not exceed 4 inches in depth; Georgia advises cultivating no deeper than 1½ inches. Where clean cultivation is used, the establishment of annual winter cover crops in alleyways is encouraged to reduce erosion, build soil organic matter, and act as a catch crop for nutrients. Establishment of a good cover crop in late summer or early fall also precludes the need for fall cultivation, which stimulates excessive late-season cane growth—tissue that is especially vulnerable to winter kill.

Figuer 3. Parts of a Bramble plant

Permanent sodded alleyways offer the greatest protection against soil erosion and the loss of nutrients through leaching, while also suppressing weeds. Conventional recommendations for sodded alleyways advise the maintenance of a four-foot-wide vegetation-free strip in the plant row, to reduce the competitive effects of the cover crop on the berries. The selection of plant species for use in sodded alleyways is also important. Non-aggressive sod-forming grasses like bluegrass, fescue, and perennial ryegrass (*Lolium perenne* L.) are often recommended, as these are only moderately competitive with brambles. Aggressive species like bermuda are definitely discouraged. The use of non-aggressive legumes as winter cover crops or as a companion to grasses in sodded alleyways offers the advantage of providing a significant amount of the annual nitrogen needs through biological fixation. The downside of legumes is the buildup of stinkbug and tarnished plant bug populations. In Oklahoma, at the Kerr Center for Sustainable Agriculture, a blend of subterranean, Dutch white, and ladino clovers was successfully used in alleyways in a blackberry planting. Frequent mowings appeared to reduce the threat from stinkbugs and plant bugs and prevented weed seed development. No evidence of winter damage from excessive soil nitrogen was observed. In a Canadian study, raspberries were grown with white clover and perennial ryegrass covers. The white clover cover resulted in higher berry yields, improved vegetative growth, and better nitrogen nutrition than the perennial ryegrass. When compared to clean cultivation, yields with the white clover cover crop were comparable. In all cases, sodded alleyways do compete with the bramble crop for nutrients and moisture. Their benefits, however, usually outweigh the negatives; growers can compensate by applying supplemental irrigation and additional nitrogen fertiliser if needed.

PRUNING & TRELLISING

The appropriate means of pruning, training, and trellising a bramble crop is largely dictated by the species and varieties grown. Good information on specific procedures is provided in conventional literature available from Cooperative Extension. Destruction of spent floricanes is important for disease and insect pest control. Removing and burning these floricanes is commonly recommended, though shredding and soil incorporation are sometimes practiced. While trellising is necessary on most trailing brambles, it is also an option for most raspberries and a number of other bramble species.

Figure 4. Bramble fruits

While somewhat expensive, trellising increases air movement in the plant canopy—reducing the need for sprays to manage plant diseases. Trellising also enhances the effectiveness of pesticides, by allowing better spray penetration. And because trellising allows easier access to fruit, cleaner picking is possible, resulting in fewer overripe berries—lessening the attraction to picnic, sap, June, and Japanese beetles. Air movement and sunlight penetration can also be enhanced by using alternate row production in regions with suitable growing seasons. In this approach, every other row is mowed during dormancy, resulting in a much more open field during the early part of the season. Aggressive regrowth in the primocane rows is exceptionally competitive with weeds, and yields in floricane rows are heavy. Reduced pruning cost is an additional benefit.

Weed Management

Aspects of weed management already discussed include suppression via cover crops and the importance of pre-plant weed control. Weeder geese can eliminate most of the grass and many of the tender broadleaf weeds from a planting. Since geese will also eat the ripe fruit and perhaps even some of the newly emerging primocanes, their use should be timed appropriately. Obviously, stocking rates will be much lower and management will be easier on clean-cultivated plantings. Investigators at the Kerr Center for Sustainable Agriculture have used weeder geese for effective weed control in blueberries with sodded middles. The center's strategy entails the use of movable electric fencing and intensive grazing.

MULCHING

Research over the years has demonstrated definite advantages to mulching in raspberries and thornless blackberries. Mulching conserves moisture, moderates temperature fluctuations, decreases weed pressure, and enhances yield. Results of research at the Missouri Fruit Experiment Station, however, show that while mulching with organic materials can be helpful in controlling some weeds, it is not sufficient for total weed control by itself. In fact, excessively deep in-row mulching can interfere with primocane emergence. It may also harbor damaging rodents. Mulching can also aggravate phytophthora root rot problems on susceptible red raspberry cultivars grown on heavy (clayey), irrigated soils. In such a situation, New York researchers recommend straw mulch during the establishment year only.

Geotextile mulches, such as woven plastic and weed barrier fabrics, are gaining favor as long-term weed control solutions, and appear suitable for use in bramble plantings. The fabric must be slit properly to minimise suppression of emerging primocanes. Plantings have been made using the fabric solely as an in-row cover, and also as a seamless mulch sheet, extending row to row across the middles. Though the initial cost is high, it may prove reasonable when amortised over the expected lifetime of 10 to 12 years. Available fabric mulches include Sunbelt™ by DeWitt Co. and Lumite by Shaw

Fabrics. Additionally, county or state Extension Services can provide information on fabric mulches, and, in some cases, might supply these products themselves.

Fertility Management

Sustainable fertility management in brambles begins well in advance of planting, with soil testing and site preparation. Some specific aspects of site preparation for bramble crops should be noted:

1. Brambles tolerate a fairly wide pH range from 4.5 to 7.5. However, the optimum pH range is 6.0 to 6.5. Lime or other suitable amendments should be applied in the year prior to establishment to bring the soil into this pH range.
2. Avoid excessive applications of phosphate. High levels of soil phosphorus have been associated with zinc deficiency in brambles. Growers planning to use poultry litter prior to establishment or as a maintenance fertiliser should take special note, since poultry wastes are especially high in phosphates.
3. Brambles are also sensitive to chlorine salts, and natural sources of potassium chloride should not be used either pre-plant or as a maintenance fertiliser.

Even when cover crops and mulches are used, annual applications of supplementary fertiliser—especially nitrogen—are needed in most bramble plantings to sustain good yields. Organic growers primarily use manures, composts, animal by-products, and vegetable or seed meals to make up the difference. Estimates of total nitrogen fertiliser requirements are provided in Table 1.

Table 1: Estimates of Nitrogen Requirements for Bramble Crops

Bramble Type	Growth Stage	Pounds per Acre of Nitrogen
Upright blackberries Summer-bearing	Establishment Year	25 to 30
purple & black raspberries	Maintenance	50 to 60
Trailing blackberries Summer-bearing	Establishment Year	30 to 35
red raspberries	Maintenance	65 to 75
Primocane fruiting raspberries (no summer)	Establishment Year	30 to 40
	Maintenance	95 to 115

During the establishment year, application should be delayed until the canes have emerged, especially if salt-based materials or manures are applied, since young brambles are easily salt-damaged. Maintenance rates shown can be applied as a single spring application (pre-bloom), or split to apply one-half immediately after harvest. Additional guidelines for supplemental fertilisation include:

1. Be certain to credit the nitrogen contribution from cover crops and reduce supplementary fertiliser rates accordingly.

2. It is a good rule of thumb to assume that only 50% of the nitrogen in most manure or compost will be available during the year of application. (The availability of nitrogen in fresh poultry manure may be closer to 90% in the first year.) However, the remaining nitrogen will become available in subsequent years and should be credited accordingly.

3. Be especially cautious to avoid nitrogen burn and nutrient imbalances when using manure. The fertiliser value of manure is highly variable and usually unbalanced with respect to nitrogen, phosphorus, and potassium (N-P-K). Considering these properties, specific application rates are impossible to recommend. Typically, farmers apply manures at roughly 1-4 tons per acre, depending on the fertility of the soil. Poultry manures, which are usually higher in nitrogen, are applied at 1-2 tons per acre.

4. The timing of manure applications is very important. Manure cannot be applied to the plants within 120 days of harvest under conditions where the fruit is either in contact with the soil or can be splashed with soil from rainfall or irrigation. Manure cannot be applied within 90 days of harvest where the fruit is elevated or otherwise shielded from soil contact.

5. If manure is fully and properly composted, it can safely be applied at higher rates and at any time during the season. For further details on manure and composting requirements.

6. When using organic fertilisers—especially unprocessed manures—growers find that any material not finely textured is difficult to spread within an established bramble hedge-row. Large clumps of organic matter can also impede primocane emergence. Coarse-textured compost or manure is more easily applied if sieved or simply broken up with spading forks or shovels.

7. In contrast to animal manures, cottonseed meal (N-P-K approximately equal to 7-2-2) is more predictable as to nutrient content, and its texture makes it easy to apply to brambles. Though more expensive than cottonseed meal, blood meal (N-P-K approximately equal to 12-1.3-0.7) is widely available, easily applied, and more rapidly available to plants. To apply roughly 60 lbs of actual nitrogen per acre, one would need to apply almost 860 lbs of cottonseed meal or 500 lbs of blood meal per acre. This is equivalent to about 2 1/3 lbs of cottonseed or 1 1/3 lbs of blood meal per 10 feet of row (assuming 12-foot row spacing).

8. All solid organic fertilisers should be applied in the mid- to late winter to allow adequate decomposition time before the spring growth spurt. Also, if applied too late, solid organic fertilisers may still be releasing nutrients to plants late in the season, which can result in cold damage to tender late growth.

9. Magnesium and boron deficiencies sometimes occur in brambles, though these are less likely under a sound organic fertilisation scheme. Magnesium deficiencies can occur under conditions where excessively high potassium levels are accompanied by low calcium. Dolomitic limestone, sul-po-mag, and epsom salts are suitable for correcting soil deficiencies. Boron deficiencies are most likely in high pH soils, particularly in the Northwest. Small amounts of Solubor™ or borax are commonly used as amendments.

10. Biannual soil and/or plant-tissue testing is advisable to identify nutrient imbalances and adjust fertilisation accordingly. Testing services are usually available through county and state Cooperative Extension.

Diseases

The foundations of organic disease control are sanitation (removal and destruction of infected plant parts); keeping the plants in good vigor through weed control, good fertility, and moisture management (mulching and irrigation); and using resistant varieties. For descriptions and diagnoses of plant diseases, refer to Cooperative Extension publications or contact an Extension specialist.

PHYTOPHTHORA ROOT ROT

Phytophthora root rot is not known to be a problem on blackberries and causes minimal problems in black and purple raspberries. It can, however, be a problem in some soils on susceptible varieties of red raspberries. A listing of known resistant and highly susceptible varieties, compiled from several sources, is provided in Table 2.

Table 2: Phytophthora Resistance of Raspberry Varieties

Phytophthora–Resistant Varieties		Phytophthora–Susceptible Varieties	
Boyne	Latham	Canby	Munger
Bristol	Lauren	Chilcotin	Reveille
Cherokee	Meeker	Comox	Ruby
Chilliwack	Newburgh	Cumberland	Skeena
Dundee	Nordic	Festival	Taylor
Fall Red	Pathfinder	Heritage	Titan
Killarney	Sumner	Hilton	Willamette
Jewel	Sunrise		

Soils naturally suppressive of phytophthora fungi contain large amounts of organic matter, plentiful calcium, and nitrogen in the ammonium form (decaying organic matter will produce ammonium nitrogen). Planting on raised beds is also helpful. Cornell researchers found that amending the soil with gypsum (calcium sulfate) reduced the incidence of phytophthora in red raspberries. Other amendments tested—including compost, lime, sulfur, and potassium—were either ineffective or inferior to gypsum against phytophthora.

Verticillium Wilt

Verticillium wilt is caused by the fungus *Verticillium albo-atrum*. To avoid soil-borne verticillium wilt problems, an often-repeated recommendation is that growers should wait at least three to four years before planting in a field where potatoes, tomatoes, peppers, eggplant, or tobacco have been grown; however, research has shown *Verticillium* species to be capable of surviving in the soil for more than 14 years in the absence of a host. Also, common weeds such as black nightshade, redroot pigweed, lambsquarter, and horsenettle support verticillium. Soil solarisation can greatly reduce the verticillium inoculum. Among red raspberries, which tend to be more tolerant, Cuthbert and Syracuse have shown to be resistant under field conditions. Among blackberries, Himalaya and Evergreen are verticillium-resistant.

CROWN GALL

Crown gall is caused by the bacterium *Agrobacterium radiobacter* var. tumefaciens, which usually enters through wounds in the roots or crowns of raspberry and blackberry plants. The disease is best controlled by planting healthy stock on a "clean" site. On a planting site where crown gall has been known to occur, a waiting period of three to five years is advised before replanting. As an alternative, a biological treatment (dip) with a non-pathogenic strain of a closely related bacterium (*A. radiobacter* var. radiobacter strain K84) can protect the planting stock from infection by the virulent *A. radiobacter* var. tumefaciens. Two commercial formulations of this treatment are available.

GRAY MOLD

The only serious disease of bramble fruit is gray mold, caused by the fungus *Botrytis cinerea*. Gray mold can be devastating if rainy weather coincides with harvest when fruits are at their ripest and most susceptible. Raspberries are more susceptible than blackberries. No organically acceptable fungicides are available for controlling this disease, but there are biological controls proven effective against gray mold. Unfortunately, none of these biological controls are as yet EPA-registered for use on aboveground plant parts. The incidence of gray mold can be reduced through advanced planning in site selection—choosing a location with good soil and air drainage, and orienting crop rows with prevailing breezes. Subsequent cultural management to improve sunlight entry and air movement (e.g., trellising, alternate row production, removal of spent floricanes, thinning, weed control, and cover crop mowing, etc.) and maintaining a tight picking schedule to reduce the presence of overripe fruit also contribute greatly to suppressing gray mold.

LEAF SPOT

Another fungal problem in humid climates is leaf spot (causal organism: *Sphaerulina rubi*). Most blackberries are resistant, but raspberries—especially those growing in the southern extremes of their adapted region—can be hard hit. Practicing good sanitation, especially the removal and destruction of old floricanes, and managing to increase air movement and sunlight penetration work to suppress leaf spot. Overwintering inoculum can also be reduced with a single delayed-dormant spray of lime sulfur. There is some genetic resistance to leaf spot among raspberry varieties. Heritage, Southland, Fall Gold, and MN 659 appear especially resistant, and black raspberry varieties demonstrate moderate resistance. Red raspberry varieties Sentry, Taylor, Skeena, Killarney, and Canby appear most susceptible.

ORANGE RUST

Orange rust affects blackberries and black raspberries. The organism *Gymnoconia peckianus*, which causes rust-colored lesions on leaves, will commonly infect the whole plant, stunting some plants seriously. Control measures largely focus on the planting of clean stock and diligent scouting to rogue out infected crop and wild plants, roots and all. Practices to improve air circulation are also advised. Among blackberry cultivars, Eldorado, Raven, Snyder, and Ebony King appear resistant. Among thornless blackberry varieties, Hull, Chester, and Arapaho appear resistant.

ANTHRACNOSE

Anthracnose (causal organism: *Elsinoe veneta*) can be a serious disease on blackberries and black raspberries. It is less of a threat to purple and red raspberries, and most of the thornless blackberry varieties exhibit resistance. Again, sanitation procedures and practices to improve air circulation play major roles in disease suppression. Anthracnose can also be controlled organically with a single delayed-dormant spray of liquid lime-sulfur. Under severe conditions, additional later sprays using bordeaux, burgundy mix, or a fixed copper may help.

SPUR BLIGHT

Spur blight (causal organism: *Didymella applanata*) is a fungal disease of red raspberries and sometimes of purple raspberries that can reduce the yields and life of a planting. Control procedures are essentially the same as those for anthracnose. The cultivars Amity, Festival, Haida, and Prestige all show some resistance.

VIRUSES

There are several bramble viruses of economic importance. Again, planting clean stock and rogueing out and destroying infected plants are important practices in limiting virus spread. Raspberries can be particularly hard hit in areas where aphid vectors of viruses are present. Breeding for resistance to aphid vectors has proved highly effective in restricting virus infection in plantings.

INSECTS

Aphids

Besides vectoring viruses, aphids can be troublesome on brambles when feeding is so intense that leaves become curled and distorted. High aphid populations are often an indication of excessive nitrogen fertilisation—a more frequent problem where soluble nitrate fertiliser is used. Organic growers usually find that, if they don't spray, naturally occurring enemies of the aphids will eventually exert control. However, organic growers may also use insecticides such as soap sprays, rotenone, and pyrethrum for control.

Borers

Other major insect pests of brambles include the borers: the raspberry crown borer, the raspberry cane borer, and the red-necked cane borer. Despite the names of these pests, all three attack blackberries as well as raspberries. Control of all three is best achieved organically by rogueing out and destroying infested plants.

Beetles

Japanese and June beetles can be major pests of bramble fruit if their adult emergence coincides with the ripening. Since ripe berries are picked every day or every other day, there are few pesticides (organic or otherwise) that are both effective against these beetles and usable up to the day of harvest. Some botanicals—such as rotenone—can legally be used even the day of harvest according to current label restrictions; however, none have proved adequate for June and Japanese beetle control. Consequently, organic growers have to resort to other methods to control these pests. Hand picking, trapping, exclusion with row covers, and reducing the immatures (grubs) in the soil with tillage, milky spore disease, and/or beneficial nematodes have all been attempted by growers with varying degrees of success. Targeting the grubs requires advance planning—the beneficial nematodes and milky spore disease are not effective against adult beetles. Grubs can be especially plentiful in undisturbed pasture or turf soils. Tillage and soil treatment with beneficial nematodes or milky spore disease are helpful in destroying pupae or grubs, but since the adults can fly in from relatively distant sites, it is often impractical to till or treat enough ground adjacent to the bramble planting to effectively suppress a local population.

Attempts have been made to exploit a behavioural phenomenon of these pests for control. Apparently the beetles are initially attracted by the odors of ripening fruit and fruit yeasts. After the first beetles find the food source, they emit an aggregation pheromone which functions as a scent beacon for other beetles. Commercial traps that use these pheromones for trapping the beetles are available through mail-order garden supply companies, but growers and researchers alike caution that they can end up attracting more beetles to the planting than the traps can handle, making the problem even worse. However, understanding this behavioural phenomenon reinforces the importance of hand picking or otherwise dealing with those first few beetles.

Arkansas researchers report that buckets of water and rotting fruit attract June beetles. The water can be laced with a pesticide or the beetles scooped off and otherwise killed. In any case, these researchers caution that the buckets must be emptied of beetles regularly because the pests soon accumulate to the extent that the last-arriving beetles may be unable to reach the water and thus easily crawl out. The fruit brew may also need to be changed if it appears to have lost its yeast-like smell or is no longer attracting the beetles. In all instances where beetle traps are employed, current wisdom advises placing them away from the field to avoid the possibility of drawing the pests to the crop. Surround™ WP is a kaolin-clay-based insect repellent effective against both June and Japanese beetles. It is considered "organic" by the Organic Materials Review Institute, and is registered on brambles. However, the clay is nearly impossible to wash off from between the drupelets (the individual seed-bearing structures that together make up the "berry," which, botanically, is more correctly called a "drupe").

Refugia

While cover crops may provide shelter to several bramble pests, cover crops and adjacent vegetation also harbor beneficial insects that pollinate the crop and help to suppress pest insect and mite populations. When crops and field borders are managed with beneficials in mind they are often referred to as refugia, and represent a new approach to pest management based on planned biodiversity.

ECONOMICS AND MARKETING

In the U.S., large-scale commercial bramble production is located almost exclusively in states along the Pacific Coast. The remaining bramble fruit production is scattered in small plantings throughout the United States. Most of the commercial crop is processed, with only a small amount making its way to the fresh market. In general, fresh bramble fruit doesn't ship well and thus is generally most appropriate for local markets. Still, it can bring high prices. More than 95 percent of the bramble fruit grown in Washington and Oregon is sold for processing. California brambles are grown mainly for the fresh market, since shippers use the fresh-market infrastructure developed for strawberries to handle and sell raspberries. Statistics are not available for other states, but in California the fresh market, especially direct-to-consumer sales, reportedly accounts for most sales.

Demand for both fresh and processed bramble fruit is increasing. Recent reports indicate that prices and sales are rising. Raspberries are the most important commercial bramble fruit, and the black and red types the most popular. Purple and yellow raspberries are less widely grown than the others, but demand is still strong. Blackberry acreage and yield are also increasing as that crop becomes more popular. Bramble fruits can be a good crop to add to the production mix for the small-scale and/or part-time farmer. Once established, raspberry plantings, for example, should produce for at least six years, and some produce for more than 20 years. Bramble fruits, which ripen shortly after strawberries or later in August or September, can extend the small producer's harvest season. However, since bramble fruits have special production requirements and a very short shelf life and marketing season, growing them is not appropriate for everyone.

The high cost of machinery, irrigation, quality plants, and labor requirements often discourages growers from getting into the bramble market. Economically, raspberries or blackberries are considered to be a medium- to high-risk crop because of a high initial investment, returns delayed for two or more years, biological factors including the climate, and high fixed costs. Establishment costs can range from $3,000 to $4,500 per acre. Plantations should consist of at least 3 acres to provide enough fruit for sale. Most growers report needing at least five to ten good pickers per acre to keep up with the ripening fruit. Growers who plan to harvest their crop with mechanical harvesters generally need at least 10 acres in production to justify renting a machine. In budgetary information developed by Kansas State University in 1993, the total variable and fixed costs for an acre of red raspberries were $2,222 in the establishment year and $1,720 in each of the three subsequent years. Based on these costs, it was estimated that, at an average yield of 1 ton per acre and a market price of $1.25 per lb, the grower would remain in a negative cash position for at least three years, since no crop is produced in the establishment year.

However, market prices will vary by region, affecting economic returns. For example, while the Kansas study above was based on a market price of $1.25, a survey conducted by Ohio Sate University in 2001 showed average black raspberry prices in Ohio to be $2.37 per lb in pick-your-own markets, $3.33 per pint in on-farm markets, and $3.79 per pint in farmers' markets. While the bulk of brambles are grown conventionally, there is some commercial organic production. Organic production is typically more costly than conventional production. Among the factors that often increase per-unit production costs in organic systems are the added labor and machinery costs associated with weed control, the loss of benefits associated with scale (organic operations are typically smaller), and reduced yields due to pest pressure. However, small-acreage organic raspberry fields, carefully hand-harvested every day over a 4-week period, can be lucrative. Expected yields in 1997 for mechanically harvested organic raspberries (in the Pacific Northwest) were between 3 and 4 tons an acre, with hand-harvested berries yielding 4 to 5 tons per acre. Prices have been high, with premiums of up to 35 cents a pound over conventional. However, increased acreage in both the U.S. and overseas is expected to continue to drive down prices by increasing supply.

Marketing

Can you make money growing raspberries? To help decide, first consider where to market the berries. Options include wholesalers, cooperatives, local retailers, roadside stands, pick-your-own, farmers' markets, and processing firms. Many small, independent fruit producers find it increasingly difficult to market their berries. Eighty percent of all fresh fruits and vegetables sold at the retail level are presently handled by chain stores in the United States. High labor costs and lack of a wholesale market at a fair return has made direct marketing much more attractive to small growers. The cost/price squeeze is particularly dramatic in the case of brambles. Direct marketing provides a means of increasing the grower's net return.

PYO marketing is highly attractive as a low-investment alternative. However, researchers found that these operations were required to sell at relatively low prices, PYO being less popular today than it used to be. Also, marketing options for PYO can be limited for operations located far from population centers. Inclement weather can have a particularly depressing effect on this marketing strategy, too. Hand-harvested berries can be sold directly to consumers or to retail. Hand-harvesting requires the management of additional seasonal workers. The researchers found that growers need to charge at least $1.00 more per pound for hand-harvested berries than PYO-harvested berries to turn a profit. Machine-harvested berries are generally taken to a processor for freezing or juicing. Low yields and few hours of machine use per year increase the cost of machine-harvested berries. Mechanical harvest requires a business that can handle large volumes of berries in one day. Processed berry prices tend to change more rapidly than fresh berry prices. The researchers found that when growers with small farms are capable of producing an average of 3,200 pounds or more per acre, they are able to compete with larger farms in marketing machine-harvested berries. However, it may be difficult to locate processors outside the Pacific Northwest.

Marketing Health through Bramble Fruits

The demand for nutritionally enhanced foods continues to grow. Beyond their vitamin and mineral content, bramble fruits are now known to contain important phytochemicals (naturally occurring compounds). The term *nutraceutical* refers to a phytochemical substance that may be considered food (or part of a food) and that provides medical or health benefits, including prevention and treatment of disease. Nutraceuticals represent one of the most important food-industry trends for marketing to an increasingly knowledgeable and health-conscious consumer. Information on nutraceutical properties of bramble fruits can be found at The Ohio State University Extension.

Value Adding

Since bramble fruits are so perishable and easily damaged, the farmer may have a lot of damaged berries unsuitable for the fresh market on his or her hands. While it is not recommended to rely exclusively on culls for value-added products, culls can make up a large part of fruit ingredients in products such as fruit leathers and jams and jellies.

REFERENCES

Anon. 1991. "Boron for brambles and blueberries". *Northland Berry News*. March. p. 6.

Daar, Sheila. 1989. "Trichoderma biofungicide registered in U.S". *The IPM Practitioner*. September. p. 19.

Edberg, Kevin. 1994. "Alternate row cropping of Boyne raspberries". *Northland Berry News*. September. p. 16.

Goulart, Barbara. 1991. "Raspberry trellises: Pitfalls and possibilities". *Northland Berry News*. June. p. 14.

Himelrick, David G. 1994. "Raspberry root rot". *American Fruit Grower*. June. p. 16.

Jennings, D.L. 1988. *Raspberries and Blackberries: Their Breeding, Diseases and Growth*. Academic Press, New York, NY. p. 87, 88.

Kleiner, Bill. 1993. "Understanding orange rust on black raspberry". *Northland Berry News*. June. p. 14.

OSU. No date. "Brambles—Production Management and Marketing". *Ohio State University Extension Service. Bulletin* 782-99.

Pritts, Marvin P. 1991. "Mulch those newly planted raspberries". *Organic Farmer*. Summer. p. 38-39.

Schaefers, George A. 1988. "Studying aphid-resistant brambles". *Fruit Grower*. June. p. 8.

Wiffig, Hans. 1997. "Organic red raspberry production in the PNW". *Northland Berry News*. Vol. 11, No. 2.

11

Organic Breeding of Garlic

Garlic, *Allium sativum*, is a member of the lily family. It may be divided into two subspecies: *Allium ophioscorodon* (bolting or hard-neck cultivars) and *Allium sativum* (non-bolting or soft-neck cultivars). *Allium ophioscorodon* produces elongated flower stalks, often referred to as scapes, and flower-like bulbils at the top of the stalk. Soft-neck garlic does not produce bulbils except in times of stress. While both bulbils and individual cloves can be propagated vegetatively, bulbils take longer—up to two seasons—to produce mature bulbs, and require special care because the young plants are very small and fragile. Hard-neck cultivars like Rocambole and Porcelain usually do better in colder climates. The cloves are larger and easier to peel. A few of these variety names are 'Spanish Roja,' 'German Red,' 'Carpathian,' and 'Music.'

Soft-neck garlic cultivars (Silverskin or Artichoke) are not recommended for northern climates. Numerous strains exist, having been selected over the years by the various companies that produce them for dehydration, or by growers producing them for the fresh market. Mechanized farms grow and develop cultivars of soft-neck garlic because the planting process can be mechanized. Since they don't produce a scape (flower stalk), the cloves can be planted upside down. Topsetting (hard-neck) garlic cloves must be set upright. There are fewer cultivars of soft-neck garlic, compared to topsetting. The varieties 'California Early' and 'California Late' comprise 90 percent of the soft-neck types grown commercially.

Garlic is well adapted for production in most parts of the temperate U.S. Yield and quality vary with climate, region, altitude, soil and pH, cultural practices, and variety of garlic. The term "biological elasticity" describes garlic's ability to acclimatize to these factors over time. That is, the same variety can look, taste, or produce differently depending on its geography, altitude, etc. No one practice is best suited for every situation. You will want to talk with other garlic growers in your area and experiment with different cultural practices and varieties to discover the best combination for your operation.

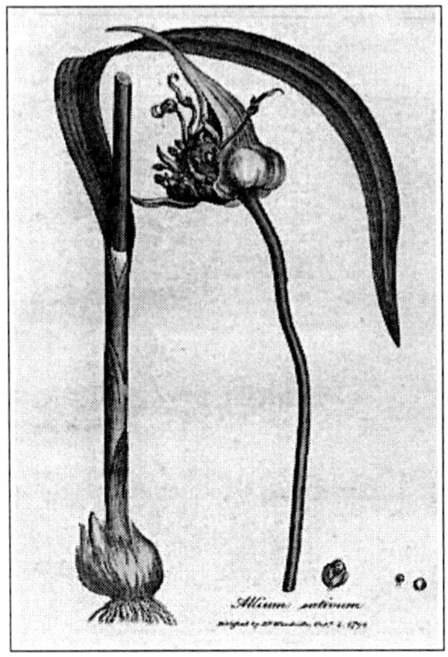

Allium sativum

Certain considerations and practices in garlic production will be the same for both organic growers and conventional growers within a given region. For instance, site selection and planting techniques are similar for both conventional and organic garlic culture. Information on these topics is available through the Cooperative Extension Service, the garlic growers associations, and common vegetable and herb texts, bulletins, and trade magazines.

Although garlic is listed in many commercial seed catalogs, "Buy local!" is the advice from David Stern of the Garlic Seed Foundation. If you buy from out of state, it will take a couple of years for the garlic to acclimatize itself, due to the biological elasticity mentioned above. David says there is an ongoing debate about varieties, and recent research with a northeastern SARE project is working to clarify this.

Many farmers will initially source their garlic seed from one of the sources listed above and then sell only what they do not intend to plant for the following season. This assures that the garlic seed will become adapted to their particular climatic and soil conditions. Other growers prefer to purchase new seed every year, as they are able to obtain high prices for their garlic and it is much cheaper to just buy the seed.

SOIL FERTILITY

We recommend getting a soil test before you begin field preparation. Request recommendations for nutrient requirements for onions when you send a soil sample to a soil-testing laboratory. Many soil testing laboratories only include the conventional NPK recommendations.

Since garlic is a high-value crop and a heavy feeder, it deserves your best ground. It needs full sun and a full range of available nutrients. A pH of 6.8 to 7.2 is ideal; many nutrients are tied up in soils that are more alkaline or more acidic than this. Garlic will grow in almost any well-drained, friable (easily crumbled in the hand) soil, preferably with high organic matter content. High organic matter aids in soil water-holding capacity and drainage. If possible, begin soil preparation the year before planting. In his book *Growing Great Garlic,* Ron Engeland recommends building up the soil over a period of 1-2 years using animal and green manures before the garlic is planted.

Provide additional nitrogen, if needed, through supplemental use of organic fertilizers. Nitrogen can be applied in the fall at planting if a slow-release fertilizer such as soybean meal is used. Avoid applying any form of soluble nitrogen fertilizer in the fall to prevent contamination of ground water as well as loss of nitrogen to leaching. Do not apply nitrogen when the bulbs are beginning to enlarge, since it will encourage excessive leaf growth and reduce bulb size. Another way to add fertility is to side-dress with compost after leaf emergence in the fall, then apply fertilizer again in the spring. Avoid fertilizing beyond May, since high nitrogen levels at this stage may actually decrease bulb size. Some organic growers apply foliar sprays of liquid fish and seaweed fertilizer, several times in the spring.

When to Plant

Fall/early winter planting is recommended in all parts of the U.S. A cold period is required in order for the mother bulb to split into cloves. The bolting cultivars of garlic require a cold period to trigger cellular division. For spring planting, which is not recommended, the bulbs need to be refrigerated at 40°F for 40 days. By planting garlic in the fall, the plants obtain significant root growth before the ground freezes. Then in the spring the plant focuses its energy on sprouting, leaf growth, and ultimately bulb development. If the garlic is planted too late in the fall to obtain significant root growth, this will ultimately detract from bulb size the following summer. Fall-planted garlic grows rapidly when the weather warms in spring.

Bulb growth in garlic, like many alliums, is dependent on the lengthening of the day and the accumulation of degree days (heat units). In northern latitudes, most growers plant garlic in October before the ground freezes. This gives the plant time to make good root development but not enough time to make leaf growth. Where winter sets in earlier, growers are recommended to plant garlic two to three weeks after the first frost (below 32°F).Where winters are milder, garlic can be planted as late as mid-December. In some parts of California, it can be planted as late as February or March. We recommend talking with local producers and your county Cooperative Extension to determine the best time to plant garlic in your region.

How to Plant

Garlic is propagated vegetatively from the clove. One clove produces one bulb. The size of both the clove and the bulb is an important consideration when selecting planting stock. Grade your garlic for both size and quality. Bulbs should be firm without soft spots or defects and the cloves creamy white without any discolorations. Discard anything that appears diseased, small, soft, damaged, or discolored. This is time-consuming, but important.

Crack each bulb into individual cloves (referred to as "popping"). Plant cloves with the root end down. Where winters are mild, plant cloves one inch deep; where winters are severe, put them two to four inches deep. In general, plant the biggest and best cloves from the best bulbs. They will generate the biggest bulbs. This will also build up bigger seed stock over time. Keith Stewart of Keith's Farm plants the medium to large bulbs, refraining from the largest. "It should also be said that planting the biggest cloves does not necessarily assure the best garlic, in terms of flavor and hardiness."

Mulch is used frequently in organic garlic production. It helps improve winter survival, suppress weeds, conserve soil moisture, and prevent soil erosion. It can also increase yields by keeping the soil cooler. Garlic quits growing when the soil temperature increases to above 90°F. David Stern establishes a winter cover crop/mulch on his New York state farm by planting oats in late August or early September. The oats are six inches tall when he plants garlic. He uses a disc-furrower on a tool bar behind his tractor to cut slices through the oats. Garlic is set into the furrows. The oats continue to grow until killed by winter frost. This system works well in areas with deep freezes, however, in milder climates, the oats will become a problem the subsequent spring if they are not winter-killed. The oats go from a living mulch in the field to a dead mulch in the winter, when the erosion potential is strong. No "living-mulch" crop combinations have been identified that do not compromise yields in garlic.

The garlic will have no trouble pushing through an inch or more of mulch. Mulch is typically used by operations without mechanical harvest, as it will make harvesting by machine difficult or impossible.Garlic is often planted in raised beds for ease of digging, good soil drainage, and reduction of soil compaction. Hood River Garlic uses a bed maker that is attached to their tractor. Terry Browne Hixson, co-owner of Hood River Garlic, claims that "the most vital piece of equipment that we own is our John Deere Tractor. Without the tractor, tiller and raised-bed maker our crop yields could not be so successful."

Clove spacing of 6 inches by 12 inches is best, except for some varieties of the Porcelain cultivars and elephant garlic, which require a spacing of 12 inches by 12 inches to produce the largest bulbs.

SEED STOCKS

Hard-neck varieties put up a tall, woody, round, flowering stalk, or scape, that grows bulbils at the top. If the plant is allowed to put its energy into these bulbils, the bulb forming below the ground can be 1/3 smaller than if the scape is removed. Many farmers cut the scapes when they are young and tender, to eat or sell them. They can be used for pickles, pesto, or stir-fry. "Depending on the planting density, 300—800 pounds of scapes per acre can be harvested, which sell for between $1.00 and $3.00 per pound," says Stern. He advises cutting or snapping off the scapes on a sunny day so that the wound will heal/ dry quickly. If you are not selling the scapes, they should still be removed from the field as a sanitary measure to lessen the potential for disease.

IRRIGATION

Garlic needs a continuous supply of moisture when it is actively growing. If the ground is dry, one inch of rainfall per week or the equivalent in irrigation is needed for best garlic growth. Drip or trickle irrigation is recommended. If you are using mulch, lay the drip irrigation lines after planting and before mulching in the fall. Stop irrigating at least two weeks prior to harvest.

LABOR

Labor needs vary seasonally. David Stern notes that 20 to 25 "human interventions" are needed between the time he starts and the time he sells bulbs. Planting and harvest are critical times. If you do not use mulch, hand cultivation should also be a consideration in labor. You should keep this in mind if you are considering expansion of your garlic planting. If you double your acreage, you will also double the time spent on each process, such as cracking the "seed" bulbs into cloves, digging, bunching, and cleaning. Storage space needs are also doubled. Mr. Stern cautions that a garlic grower makes a large investment in this crop: tasks must be done on time or the crop and the investment will be lost. If you are planning to significantly expand your acreage, mechanizing the planting and harvesting can alleviate some of the labor requirements associated with the expansion. Erme, a Dutch company, manufactures bulb planters. A typical 12 row planter can plant 25 acres of bulbs in a day. There are also 3-, 4-, and 8-row configured planters. Erme also manufactures single-row garlic harvesters.

PEST MANAGEMENT

Healthy soil builds healthy plants and is a foundation of ecological pest management. Practices such as crop rotation and the creation of habitat for beneficial organisms work to manage pests by increasing the biological diversity on the farm.

Most pests that attack onions will also attack garlic. These include onion thrips, onion maggots, grasshoppers, and gophers. Diseases are caused by a number of fungi, bacteria, nematodes, and viruses. Learning to identify the symptoms is the first step in management. *Growing Great Garlic* has a section on pests and diseases that gives good descriptions and management techniques. Two other excellent resources are *Compendium of Onion and Garlic Diseases* and *Integrated Pest Management for Onions.* The Cooperative Extension Service can also help in pest identification.

Soil solarization is a process that uses heat from the sun for controlling many soil-borne plant pathogens, weeds, and nematodes. Basically, when clean-tilled soil is covered with clear plastic for several weeks during the hottest part of the summer, solar heat builds up under the plastic, killing many weed seeds and disease-causing organisms. This timing corresponds particularly well as a pre-planting technique for fall-planted garlic. The bulletin *Soil Solarization: A Non-Chemical Method for Controlling Diseases and Pests* gives more specific details and is available from the University of California Department of Agriculture and Natural Resources. Information on how to obtain this is available under Further Resources.

Terry Browne Hixson from Hood River Organic Garlic suggests scouting as a pest management tool. Her partner Eric spends countless hours out in his fields, especially during the summer months, when he spends many hours daily inspecting the crops.

INSECTS

Onion thrips begin to migrate when weeds in surrounding fields begin to dry up. Monitoring thrips with hot pink sticky traps should start before this migration begins. Since adults and nymphs will inhabit weedy areas surrounding the field, keep these areas weed-free to reduce thrips infestations. After the crop is harvested, the tops should be raked together and burned to reduce overwintering populations.

Several species of lady beetles, as well as the minute pirate bug, are predators of onion thrips. The predatory mites *Amblyseius cucumeris* and *A. barkeri* are also reported to attack onion thrips. Adult female predatory mites (*Neoseiulus*) consume from 1 to 10 young thrips per day and have a 30-day lifespan. They can also survive on pollen and other spider mites in the absence of thrips. Repeat applications of predators must be made to establish a 1:2 ratio of predators to prey. *Neoseiulus* attacks first instar (very young) thrips only and does not move long distances from where it is first placed. They are most often applied in small piles at the base of plants, or in paper bags. Usually, a small hole is made in the bag, and mites move out of the bag slowly.

Carol Glenister of IPM Laboratories in Locke, New York, has researched the use of predatory mites for control of onion thrips in the field. She says the results were not encouraging. Thrips are hard to control through biological controls because of their mobility and numbers. They migrate in large numbers (in the billions) from neighboring areas that are drying up, and the mites find it difficult to prey on them all at once. Glenister suggests that release of beneficial insects is not the answer, but ensuring natural habitats, and providing crop diversity would be more effective.

There are several "least-toxic," or biorational, pesticides that may reduce thrips populations. According to Dr. Richard Lindquist, entomologist at Ohio State University, a mixture of M-Pede® (insecticidal soap) and SunSpray Ultra-Fine Spray Oil® controlled western flower thrips—a closely related thrips species—in a greenhouse trial.(Lindquist, 2001) Sunspray is no longer available from the manufacturer; however Safe-T-Side horticultural oil may be an apt substitute.

Neem oil, a botanical pesticide extracted from the neem tree, *Azadirachta indica*, has shown good control of a variety of agricultural pests. Dr. Lindquist indicates that it provided control of flower thrips in the greenhouse after a series of four weekly applications. (Note: Neem is not a good rescue treatment, but works as an insect growth regulator and should be applied early in the crop cycle.) Before spraying a new biorational pesticide, always check with your certifier to confirm that the product is approved for organic production.

Onion Maggot

Beneficial nematodes have shown to be effective with onion maggots. *Steinernema carposporae* and *Steinernema feltiae* are the most effective species for maggot control, with the latter being more tolerant of cool soil. For the most effective nematode biological control, spray in the morning and evening and keep soil moist. Beneficials such as ground and rove beetles, birds, and braconid wasps prey on all life stages of the maggot.

Diatomaceous earth spread around early transplants has also shown effectiveness in treating the maggots. The fine fossilized shell material has microscopic sharp edges that pierce soft bodies and causes them to dehydrate. Apply after a light rain for better effectiveness.

Good cultural practices are essential in controlling this pest. Start by rotating your crops. Don't plant garlic in the same place two years in a row. Once the pest's breeding cycle is broken by planting its host (garlic and onions) in a different spot, you can use a floating row cover to keep the flies from laying eggs on your fresh plants. Use the cover immediately after planting cloves, and insure it extends at least six inches beyond each side of the planted bed. Remove and destroy garlic debris at the end of the season.

Bulb mites and leafminers can also be significant garlic pests in California. Bulb mites damage bulbs by penetrating the outer layer of tissue and allowing rotting organisms to gain entry. This pest is most damaging when plant growth is slowed by cool, wet weather. Bulb mites can reduce plant stands, stunt plant growth, and promote rot of bulbs in storage. Management includes rotating garlic and onions with other crops, avoiding rapid rotation of garlic into areas with high residue, and insisting on clean garlic seed. Hot water treatment of seed garlic before planting may reduce mite infestation.

Damage by leafminers in garlic is of little concern unless populations become so high as to prematurely kill foliage. Close proximity to crops such as lettuce, celery, or spinach will increase the potential for damage by leafminers in garlic. Natural enemies, especially parasitic wasps, are effective at managing leafminer populations.

DISEASES

Most garlic diseases are either soil- or seed-borne and usually can be controlled with proper rotation and by planting disease-free seed. Certain cultural practices can help to reduce the incidence of diseases in garlic. Crop rotations away from other species of alliums and from fields with a history of disease problems are recommended. Sanitation is another important aspect of disease control. If you see any yellowing or misshapen leaves in the field, they should be removed and destroyed. Culls and diseased foliage should also be removed and either burned or buried away from the field after harvest. Purchasing disease-free stock is essential in avoiding disease problems.

The most common diseases in garlic include:

White Rot

Sclerotium cepivorum is a major disease of commercial garlic grown in California and the Pacific Northwest. The organism is most active when the temperature is cool (less than 75°F). In northern climates it usually attacks in the spring. Symptoms include premature yellowing and dying of older leaves, stunting, and leaf tipburn, followed by destruction of the root system, shoot dieback, and rotting of the bulb. Control by rotating out of allium crops for many years (white rot has been known to persist in soil for 10 years), destroying infected tissue, and planting disease-free seed stock. Spraying garlic extract in the absence of the allium crop may stimulate the fungal mycelium to grow and exhaust itself several weeks before planting (soil temperature needs to be between 60 and 70°F) resulting in a successful subsequent garlic crop. Apply the garlic extract in 10 to 150 gallons of water per acre. Use of this product before planting causes sclerotia in soil to germinate and die before the crop is planted.

Fusarium (Basal or Bottom Rot)

Fusarium oxysporum and *F. cepae* are the causal organisms of this fungus, which is present in all soils and is usually considered a secondary invader because it attacks plants that have been weakened by insects, mechanical damage, or other diseases. It is the most common disease found on planting stock. *Fusarium* is most active at high temperatures. Symptoms are similar to white rot, except disease progression is much slower and death of the plant may not occur. In the field, affected plants may show reduced emergence, yellowing and/or browning (necrosis) of leaves beginning at tips. The discoloration will move toward the base of the leaf, which will eventually wither and die.

Bulbs infected with *Fusarium* may decay further in storage. This disease is controlled by proper crop rotation with non-susceptible crops for four years, removal of infected plants, and planting disease-free seed.

Botrytis

Symptoms of *Botrytis* include water-soaked stems, which is why the disease is often called "neckrot." In severe infections, the bulbs may rot. In mild infections, the disease may not be noticed during the season, but may attack the bulb during storage. Control this disease by promoting air movement through the field so that foliage does not remain wet. Rapid drying during harvest, followed by good aeration during storage, will also minimize the problem. Use disease-free planting stock.

Penicillium Molds

Penicillium is both a field and storage disease. Plants from infected cloves planted in the fall will often emerge in the spring, turn yellow, and then die. A blue-green color is observed on cloves in soil and in storage, thus its common name, "Blue Mold." When conditions are optimum for rapid garlic bulb emergence, the plant may outgrow the disease. Air-borne spores spread the disease. If a bulb is infected, do not use the cloves for planting stock. Wash hands after touching the bulb and avoid bruising or wounding stored bulbs. Prevent the disease by planting clean stock.

Rust

Puccinia allii, also called *Puccinia porri,* is the fungus that causes rust. Until recently, this fungus was considered to be of minor importance in garlic production. However, outbreaks in California in the late 1990s reduced crop yields by up to 75 percent in some fields. Initial symptoms occur on the foliage and stem as small white flecks that develop into orange spots (spores) or pustules. The bulbs become shrunken and deformed. Heavily infected plants may turn yellow and die. Conditions favorable for disease development include high humidity, low light, and low raintall and a temperature between 45 and 55°F. Disease incidence is highest in stressed plants. To reduce infection potential, use healthy seed in well-drained soil. Rotate with non-allium crops.

Other Diseases

Bacterial soft rots are characterized by softening and water soaking of one or more of the inner fleshy scales of the bulb. Affected tissue is yellow initially, turning brown as the disease progresses lengthwise in the bulb. The neck of infected bulbs may be soft when pressed. These organisms generally appear just before or at the time of harvest or in storage. This organism is more common in onion than in garlic. Wounds and senescent leaves are the means by which bacteria gain entrance into the bulb. It is spread through direct contact with water. Avoid overhead irrigation once garlic begins to bulb.

Black mold is first evident at the top or sides of the bulb where disease or injury has caused an opening in the skin. The fungus develops between dry, dead outer scales and the first inner fleshy scales of the bulb. Invaded scales initially become water soaked. Under dry conditions diseased scales dry and shrivel, and black masses of spores are visible between outer scales. Gentle handling of the bulbs post-harvest is a preventive measure for this disease.

NEMATODES

Ditylenchus dipsaci is a stem and bulb species of nematode. Plant parasitic nematodes are microscopic roundworms that live in soil and plant tissue and feed on plants by puncturing and sucking the cell contents with a needlelike mouthpart called a stylet. The stem and bulb nematode lives within the plant, feeding in stems, leaves, and bulbs. The only known hosts for the garlic strain belong to the onion family: garlic, onion, leek, and chive. The stem and bulb nematode penetrates the germinating clove and destroys tissue as it moves through, seeking food. Nematodes suck the cell contents and their salivary secretions cause the cells to collapse. Root knot nematodes can cause stunting and reduce a stand of garlic. Stubby root nematode causes stunting of plants.

Plants infested with the stem and bulb nematode have distorted and bloated tissue with a spongy appearance; the plants are stunted with shortened and thickened leaves, often with brown or yellowish spots. In the field, top growth will yellow and ripen prematurely. The whole bulb may separate from the basal plate.

Management options include:

1. Eradicate volunteer garlic plants from fields having a history of stem and bulb nematode.
2. Avoid other hosts of the nematode in your rotation if possible. To reduce reintroduction and spread of this pest, it is necessary to implement a three-year crop rotation with a non-host crop and to bury or burn infested or contaminated seed and plant debris.
3. Plant only nematode-free cloves.
4. Soak cloves at 100°F for 30 minutes in water containing 0.1 percent surfactant; then soak at 120°F for 20 minutes in the same kind of solution. Cool cloves in tap water for 10 to 20 minutes before drying for 2 hours at 100°F, or plant if possible. This helps eradicate the bulb or stem nematode from loose (cracked) cloves but not from intact bulbs.

VIRUSES

There has been a high incidence of multiple viruses found in garlic fields in Canada and throughout the United States in recent years. However, there are only rare situations where these have had economic consequences. The most obvious symptoms are "radical abnormal leaf expression, such as extreme curling, stunting, swollen veins and unusual protrusions." If your garlic crop shows these symptoms, it may grow out of it, but will still have the virus. It is important not to use seed from that crop for the following year.

WEEDS

Good weed control is *essential* in garlic production. Alliums are slow-growing, shallow-rooted crops that can suffer severe yield loss from weed competition. Planted in the fall and harvested in the mid-summer of the next year, garlic will be in the ground nine months. It is therefore vulnerable to competition from winter and summer annual weeds. Weed competition, even early in the growth of the crop, can reduce yields. In addition to reducing yield and quality, weeds also interfere with mechanical harvesting equipment.

If possible, advance field preparation with a weed-suppressing cover crop such as rye or sorghum-sudan grass can significantly reduce weeds as well as build up soil organic matter. Mulching new plantings has already been mentioned as a way to control weeds. If mulch is used, it should be thick enough to last until harvest. Avoid spent hay and inspect straw for weed seeds. If mulch is used, garlic will have to be dug by hand rather than machine harvested. Garlic can also follow any well-cultivated crop, like carrots.

Cultivation is another method of weed control. If the soil is not mulched, various types of cultivators or tillers can be used to control weeds between rows and wheel tracks on the sides of the beds. The weeds within the plant rows may be removed by hand cultivation. David Stern has adapted his tractor equipment for cultivation with small discs that throw dirt between the plants. The garlic leaves are not covered, but small weeds are eliminated.

Flame weeding works with garlic, as well as with onions. At six commercial organic farms in Europe, flaming was tried on onions before the four-leaf stage and also later in their development. Flaming the younger plants resulted in damage to the onions. However, flaming at a later stage was successful, though too much heat damaged the onions. Flaming also works best with more mature garlic plants, particularly with the Porcelain cultivars, because they have vertical leaf structure. Field conditions such as uneven ground, inconsistent tractor speed, or high winds can affect the degree of success with this technique. For small-scale plantings, backpack flamers from construction supply outlets should work well.

A SARE farm-based research grant in 2003 studied the affects of acetic acid (vinegar-based) herbicide for control of broadleaf weeds and grasses in garlic. The outcomes demonstrated that 10 percent acetic acid vinegar was very useful in controlling broadleaf weeds in the test plots. The vinegar provided minimal to no control on grasses, however.

Test plots sprayed from both sides of the row had virtually no broadleaf weeds at the end of the trial. There was no difference in plant size, maturity date, quality, yield, bulb size, appearance, or keeping ability between the test plots and the control plots. Despite the unusually cold and wet spring during which this trial was conducted (which impacted the ability to make timely applications), they found that there was a 94 percent reduction in labor (associated with weed management) by using vinegar rather than in-row weeding. It is important to use a mask and gloves when handling and spraying acetic acid-based herbicides, as they can be caustic to your skin and lungs.

HARVEST AND STORAGE

Gauging the right time to harvest is very important. Garlic will double in size during its last stage of growth. If dug too soon, the cloves will not have grown to their maximum size. If bulbs are dug too late, they may have begun to split apart in the soil. David Stern often begins harvesting when the leaf tips start to brown. Ron Engeland uses the number of green leaves left on the plant to judge if the bulb is ready for harvest. On an average, he harvests when about six plant leaves are still green. Some growers harvest when plants are 40 percent browned and 60 percent green. Mr. Stern notes that leaf conditions cannot always be an accurate indicator that it is time to harvest. Browning of leaves may be the result of drought, damage, or disease. He recommends digging some plants up to determine the correct time to harvest. The outer skin should be tight, the bulbs fully developed and well formed. Mr. Stern recommends digging sooner rather than later. If garlic becomes too mature before harvest, the cloves will begin to crack apart while still in the ground. However, if the garlic is pulled at a slightly immature stage, the leaves, which serve as wrappers for the cloves, will continue to translocate nutrients to the cloves, making the crop easier to prepare for sale.

In small-scale plantings, garlic can be dug with a garden fork. For larger acreages, several tools are available for undercutting and harvesting garlic. Bed lifters, potato diggers, or subsoilers can be used to loosen garlic from the soil. (These will not work if heavy mulch remains.) Erme, a small farm implement manufacturer from France, has single- and double-row garlic harvesters and binders.

The Garlic Seed Foundation can also provide information on building your own harvesting equipment, or referrals to manufacturers.

After mechanical digging, the garlic still needs to be removed from the field by hand. Field grading should be done immediately to remove any damaged or diseased plants, a standard practice for disease prevention.

Sort garlic into three categories:

1. Small bulbs can be cleaned for sale or your own kitchen.

2. Medium-sized bulbs are cleaned in preparation for sale. One harvesting debate involves whether or not to wash bulbs. Ron Engeland states in *Growing Great Garlic* that he doesn't want to invite mold and decay that might result from having wet bulbs as a result of washing, and bulbs grown in light soils with lots of humus clean up fairly easily. Garlic grown in heavier soils is more difficult to clean. David Stern's method for cleaning garlic begins by holding a bunch of six to eight bulbs in one hand and spraying them with the garden hose while rotating his wrist so that all sides are washed. He then lays the garlic—not bunched—on racks in the greenhouse and off the ground to dry overnight. The plants are bunched the next day, and stay bunched until he is ready to braid. David Stern digs in the morning and cleans in the afternoon, taking care not to dig more than he can clean that day. Cull any diseased bulbs right away.
3. Large bulbs are saved for planting stock. These should not be washed, but hung in a covered barn or shed to dry. Fans may be used to increase air movement in wet years. (California is the only place where garlic is dried in the field. Growers commonly lay the bulbs in the field for a few days to dry, covering the bulbs of one row with the leaves of the next to prevent direct exposure to sunlight.)

Before garlic is stored, it must be properly cured or dried. According to David Stern, the garlic needs to cure/dry for at least 10 to 14 days. The neck cells constrict and hold the juice in the bulb. He suggests this be done with the stem and roots still on, but there are many theories on this which can vary by region and year. After a couple of weeks they may be clipped off, leaving ¼ to ½ inch of the stem and roots. When the outer skins are dry and crispy, the garlic is ready for storage or sale. It can be stored in clean onion bags. Garlic will lose up to 20 percent of its weight in the curing process.

According to the publication *Commercial Storage of Fruits and Vegetables*, garlic will keep for 6 to 7 months if it is stored at 32° F and at 65 to 70 percent relative humidity. It is important to keep temperature and humidity constant. High humidity will keep the bulbs from dehydrating. A walk-in cooler should make a suitable storage facility; however many farms shut their coolers down in the winter. Also, any variation or fluctuation in these conditions will initiate sprouting. This can make it difficult to store garlic for long periods of time without sprout inhibitors. A 25-watt light bulb and a thermostat can be used to provide heat when needed. A fan will keep air circulating. Most farmers sell their garlic as a fresh crop at farmers' markets; very few sell over the winter, because of the difficulty associated with sprout prevention.

MARKETING AND ECONOMICS

U.S. garlic production has doubled over each of the last two decades. No other vegetable, including high flyers like onions, broccoli, and carrots, has exhibited such strong sustained growth. Despite this sustained growth, there has been significant global competition, mostly from the People's Republic of China. As of 2004, China was the dominant source of imported garlic in the United States, despite the imposition of a hefty duty tax against fresh Chinese garlic imports imposed in 1994. Prior to imposition of the anti-dumping duties China was a major exporter of garlic to the United States, which had a significant impact on U.S. growers. International competition should be a serious consideration for farmers considering developing larger international markets for their garlic.

The cost of seed cloves plus the hand labor for planting and harvest makes the initial investment for garlic production high in comparison to some other vegetable crops. Additional costs include land preparation and installing an irrigation system. Garlic returns are very dependent on how the crop is marketed. The University of Kentucky Cooperative Extension bulletin on garlic production suggests that wholesale marketing of well-managed garlic at prices from $2.00 to $4.00 per pound could easily return (net) from $1,400 to $3,200 per acre (Table 1). An acre of well managed conventional or organic garlic that is direct marketed at prime locations by the producer (perhaps in braids and other forms) could return in excess of $5,000 per acre. In producing value-added products such as braids, labor needs to be accounted for in your enterprise budget.

Table 1. Net Returns for five different yields and prices of garlic Yield (lbs./ acre)

Price ($/lb.)	2,000	3,000	4,000	5,000	6,000
$0.80	($5,815)	($5,313)	($4,810)	($4,308)	($3,805)
$1.60	($4,215)	(2,913)	(1,610)	(308)	995
$2.40	($2,615)	(513)	1,590	3,692	5,795
$3.20	($1,015)	1,887	4,790	7,692	10,595
$4.00	$585	4,287	7,990	11,692	15,395

Taken from: Garlic Production.(Ford et al., 2006) Penn State Agriculture Alternatives.

Of the total garlic harvested in the U.S., 75 percent is sold wholesale for dehydration; however, garlic is an excellent crop to direct market as a local specialty. It can be sold in a number of forms for a number of uses. Some fresh-market options for garlic include wholesale markets, farmers' markets, and roadside stands. Sales to local retail markets, such as supermarkets and specialty produce and health food stores, are also an option.

Retail garlic prices vary considerably. Supermarkets generally sell garlic at around $2.00 per pound, while some natural foods grocers sell garlic for as high as $4 to $5 per pound. Direct market prices can go as high as $1.00 per bulb for specialty varieties.

In addition to the whole bulbs, garlic can be sold in a number of additional forms:

1. *Garlic greens* are the leaves of the tender, immature garlic plant grown from small whole bulbs planted four inches apart in the fall. They are harvested at ground level, bunched (2" across at base), banded, and rinsed. The leaves need to be tender, and

approximately 8-10" tall. New leaves will re-grow up to two additional times. Harvest season is approximately late March to late April in most temperate climates. In warmer climates this may be earlier. You can create a value-added product by making garlic greens pesto.

2. *Garlic scapes* (or flower stalks) should be cut from the plant when tender, curled, and no longer than 16" total and/or 1/4" in diameter. Harvesting in mid-day sun is a good cultural practice. They should be rinsed, drained, bagged, and refrigerated. These are living vegetable matter and must not dry out or mold. Monitoring in storage is necessary, and yellowing scapes must be removed.

3. *Garlic Scallions* are immature plants. They can be grown from small cloves planted in either fall or spring. They should be harvested when the plant is 12" above ground and 14" total. The entire plant is dug, outer leaf or two peeled down and off, and rinsed clean. The roots should be left on and free of dirt. Pack the "scallions" for market in banded bunches of 6 with 12 bunches per poly bag or waxed and lined box. No bulbing at the base is acceptable, since this will be tough and inedible. No scallion should be less than 3/8" in diameter.

4. *Bulbs with leaves* attached may be braided, with or without dried flowers or herbs, and sold as ornamental edibles. Instructions for making garlic braids are available in the informational pamphlet titled *How to Make a Garlic Braid*, by Diane Trenhaile, listed below under Further Resources.

Many farmers sell garlic for planting stock. Planting stock can be sold to gardeners, local farmers, and through the Internet or mail order. Terry Browne Hixson, from Hood River Garlic, describes their decision to grow organic garlic for planting stock. "Seed was not our original goal. Our goal was to just grow garlic. But after years of experience and hard work, we started growing garlic of superior quality. Our markets are people who want to buy premium quality, certified organic seed garlic. We are low-volume, high-quality seed growers. We love to grow our garlic and practice good stewardship for the land. There are easier and less expensive ways to grow garlic, however, we choose to grow organically and promote sustainable agriculture to help our planet." They mainly sell their garlic via the Internet, and use advertisements in selected magazines.

Terry also suggests a Garlic of the Month club for those who are direct marketing "The Garlic of the Month Club has been a great market for us. It's a great gift idea for anyone who loves to cook, as well as the kitchen gardener."

Garlic festivals provide an excellent opportunity to market and increase consumer awareness. Almost every state has a garlic festival, check out the Garlic Seed Foundation Web site to find one near your location.

REFERENCES

Salunkhe, D.K.; Kadam, S.S. (1998). *Handbook of Vegetable Science and Technology*. Marcel Dekker.

Shuford JA, Steckelberg JM, Patel R (January 2005). "Effects of fresh garlic extract on Candida albicans biofilms". *Antimicrob. Agents Chemother.* 49 (1): 473. doi:10.1128/AAC.49.1.473.2005. PMID 15616341.

Silagy C, Neil A (1994). "Garlic as a lipid lowering agent—a meta-analysis". *J R Coll Physicians Lond* 28 (1): 39–45. PMID 8169881.

Steiner M, Lin RS (June 1998). "Changes in platelet function and susceptibility of lipoproteins to oxidation associated with administration of aged garlic extract". *J. Cardiovasc. Pharmacol.* 31 (6): 904–8.

Yeh, Y-Y., et al. (1999). Garlic extract reduces plasma concentration of homocysteine in rats rendered folic acid deficient. *FASEB Journal* 13(4): Abstract 209.12.

Yeh, Y-Y., et al. (1997). Garlic reduced plasma cholesterol in hypercholesterolemic men maintaining habitual diets. In: Ohigashi, H., et al. (eds). *Food Factors for Cancer Prevention*. Tokyo: Springer-Verlag.

12

Prunus Tree Crop Breeding

The *Prunoideae*, a subfamily of *Rosaceae*, includes several species producing edible drupes with economic importance. In 2001, worldwide annual production of *Prunoideae* exceeded 28.3 million metric tons, including almost 13.5 million tons of nectarines and peaches (*Prunus persica* (L.) Batsch), 9 million tons of plums (*Prunus domestica* L.), 2.7 million tons of apricots (*Prunus armeniaca* L.), 1.8 million tons of sour and sweet cherries (*Prunus cerasus* L. and *Prunus avium* L. respectively) and 1.3 million tons of almonds (*Prunus amygdalus* Batsch = *Prunus dulcis* (Miller) D.A. Webb) (Table 1). *Prunus* species are characterized by developing only one ovary in which two ovules typically form; one of them degenerates soon after anthesis. The fruit is a drupe where the mature stony endocarp together with the seed forms a propagation unit comparable to a botanical seed surrounded by its protective testa. Breeding practices in *Prunus* must address the challenges resulting from the narrow genetic background of commercial cultivars, a long juvenile period, and the differences in trait expression between juvenile and mature trees3. In the last decade, many techniques such as *in vitro* culture, and the use of molecular markers have become available for *Prunus* crop improvement. These methods are now moving from laboratory evaluation to field application. Simultaneously, different aspects of the utilization of DNA markers, including marker-assisted selection and genome mapping, as well as the impact of altered gene expression on *Prunus* spp., have now been extensively reviewed.

GERMPLASM IMPROVEMENT

Related Prunus species

The available germplasm in *Prunus* is diverse and the origin and dissemination for several species have been extensively reviewed. It is still possible to find considerable genetic variation for these species mainly in the mountainous areas of Central Asia from the Tian Shan region in China to Kurdistan, including Turkestan, Afghanistan, Iran and Iraq. However, in cultivated germplasm, a limited gene pool restricts production to specific areas and conditions.

The introduction of genes from related species through interspecific hybridisation has been used in several breeding programs throughout the world to develop better-adapted cultivars and rootstocks. Rootstock breeding programs using interspecific hybridisation have introduced useful traits including size control, adaptation to the new environments and pest resistance. Interspecific crosses between *Prunus* species (primarily peach × almond, but also *P. webbii* × peach, and others) have been widely utilized in almond rootstock breeding in France, USA, Spain, and Yugoslavia. The introgression of almond germplasm from related species, including *P. webbii* (Spach) Vieh., P. argentea Lam, P. persica, P. bucharica

Korshinsky, *P. mira* Koehne and *P. scoparia* Batal.has allowed transfer of several useful traits including selfcompatibility, fungal and pest resistance, and frost and drought tolerance. *P. davidiana* (Carr.) Frans. has recently been reported to be a source of plum pox virus (PPV) resistance for peach, while the introgression of *Prunus mandshurica* (Maxim.) Koehne genes to apricot have improved frost resistance in Eastern and Central European programs. In almond, the absence of extensive crossing barriers in either the initial hybridisation or subsequent backcrosses demonstrates a direct accessibility of this rich germplasm to breeding. Potential barriers to successful interspecific gene introgression include male sterility, poor germplasm maintenance, and problems associated with character quality. The encouraging performance of interspecific hybrids and backcrosses to date, support continuing opportunities for transferring useful traits, including self-compatibility, resistance to important pests and diseases, improvement of seed oil quality, tolerance to aberrant environments, and modified tree architecture and bearing habit. International collaborations have further allowed a more thorough evaluation of wild and related germplasm prior to extensive gene introgression, and has helped minimize breeding obstacles imposed by quarantine restrictions.

Multiple embryos

Twin seeds (multiple embryos within the same seed coat) occur spontaneously in several *Prunus* species including peach and almond. The occurrence of these multiple embryos varies greatly among years and is strongly influenced by environmental conditions. This phenomenon has been mainly studied in peach and in the almond cultivars 'Nonpareil' and 'Mission'. Seedlings from the same twin peach seed are frequently viable and show similar growth habits, though occasionally one of the seedlings show weak growth and develops poorly. Some of these lowvigour plants have been shown to be haploids from which truebreeding dihaploids can be generated for genetic studies, hybrid rootstock production, and transformation and regeneration studies. Some of the low-vigour, twin almond seedlings were found to be aneuploids and thus, have value for developing near isogenic lines (NIL). A collection of these haploid/aneuploid NILs are presently being developed to aid in genetic (locating genes, transfer of particular chromosomes) and molecular (isolation and sequencing of genes, genetic transformation, etc.) studies as demonstrated by Muehlbauer et al. and Young et al.

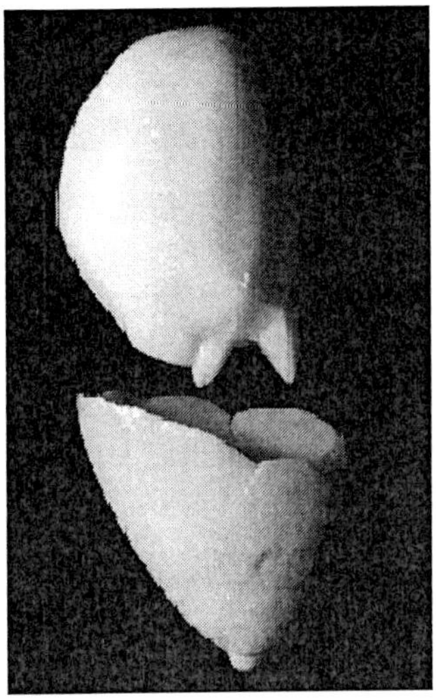

Figure 1. **Detail of twin embryos in an almond seed.**

NEW PROPAGATION TECHNIQUES

In Vitro Evaluation of Agronomic Traits

The possibility of growing plants and even isolated plant cells in a test-tube under controlled *in vitro* conditions, offers unique opportunities for improving selection efficiency. Advantages include minimizing environment influences, the potential to handle large numbers of individuals in a very small space, and accelerated growth and development. The increasing availability of other biotechnological techniques (biochemical markers, DNA analysis, genetic transformation, etc.) further complement *in vitro* culture opportunities. Wenzel and Foroughi-Wehr reported the utilization in herbaceous species of *in vitro* callus culture for the selection of resistance to environmental stress (freezing tolerance in wheat, salt tolerance in rice) and diseases (*Phytophthora* and *Fusarium* spp.), and herbicide tolerance in tobacco. The application of *in vitro* culture techniques for the selection of horticultural characters, however, may be more difficult. For temperate fruit species, tissue culture propagation has progressed rapidly during the last years. The application of tissue culture techniques as alternative propagation methods has been reported as early as the 1960s. Initially, tissue culture has involved micropropagation and somatic embryogenesis. Axillary shoot production (meristem culture) is the system most frequently utilized to regenerate plantlets by micropropagation techniques. Research in somatic embryogenesis has recently increased in anticipation of more widespread attempts at genetic transformation. Tissue culture has numerous potential uses for temperate fruit and nut tree species, including propagation of rootstocks, own-rooted scion cultivars, virus-free stock plants, and elite genotypes. These techniques offer unprecedented opportunities for the evaluation of horticultural traits in breeding programs. Applications have been reported in *Prunus* for the evaluation of the compatibility between cultivar and rootstock, the resistance to abiotic stress, and the resistance to biotic stress.

In Vivo Micrograft

Grafting has been widely used over the centuries for asexual propagation of fruit trees. Micrografts, developed in the 1970s, consist of the grafting of millimetresize vegetative meristems. Initially, this technique was used for virus elimination in fruit trees. Subsequently, it has been used for the early assessment of rootstock-scion incompatibility, commercial multiplication, virus detection and phytoplasma studies. Micrografts proved to be a useful technique when the early propagation of plant material was desired and to invigorate weak material. Optimum propagation efficiency is achieved through maximizing the different parameters involved in micropropagation and the later growth of the buds. *In vivo m*icrografting avoids tissue culture transplant shock when transplanting from sterile *in vitro* conditions. This technique has been employed to recover aneuploids of almond which occur at low frequencies in sexual embryos with seeds.

Artificial Cycle of Growth

The growth of woody plants from temperate climates requires a periodic endodormancy, which can be artificially achieved through rest periods in cold chambers. Two cycles of vegetative growth per year can be carried out by employing two 4-month periods of growth in the greenhouse and two 2-month periods of rest in cold chambers. The size of the pots, the periodic renewal of the soil and the control of mites and fungus have played a key role for the successful management of almond seedlings under these controlled conditions. The cold treatments can also be used to control fungi and mites. Seedlings, properly maintained under these conditions, can provide vigorously growing plant tissue (leaves, root tips, etc.) throughout the year. This approach has been used in virus resistance studies, as well as to maintain quarantine conditions (e.g. in the study of dangerous viruses such as the PPV causing sharka disease). Artificial cycles of growth have also been used to invigorate weak genetic material such as the aneuploid seedlings of almond. Similarly, artificial cold treatments have been successfully utilized in studies of frost tolerance in stone fruit flowers.

Use of Molecular Markers

Characterization and Genetic Relationships of Plant Material:

Traditionally, characterization and identification of *Prunus* species and cultivars has been based on morphological and physiological traits. However, such traits are not always available for analysis, and are affected by changing environmental conditions. Molecular marker technology offers several advantages over the sole use of conventional markers. Molecular markers developed for *Prunus* also offer a powerful tool to study the evolution of the genome, and for understanding of genome structure and determinants of genetic diversity.

Isoenzymes

Isoenzymes were among the first genetic markers to be widely utilized. They have been used for cultivar identification in *Prunus* because of their environmental stability, their codominant expression, and their good reproducibility. Nevertheless, their utilization is limited by the small number of loci that can be analysed with conventional enzyme staining methods as well as the low variation in some loci. Electrophoretic surveys were particularly useful in characterizing almond and plum cultivars, because both almond and plum are outcrossing species with high level of isoenzyme polymorphisms. In contrast, peach, a predominantly autogamous species, shows few isoenzyme polymorphisms in spite of its extensive morphological variability. Apricot shows intermediate levels of variability with the predominantly out-crossing non-European populations exhibiting higher isoenzyme variability than the predominantly inbreeding European populations60. Isoenzyme analysis has also been used to identify different interspecific hybrids and detect phylogenetic relationships among species64. More recently, isoenzymes in combination with DNA-based markers were employed to create the genetic maps for woody perennials and for the genetic characterization of multiple embryos in almond.

RFLPs

Restriction fragment length polymorphism (RFLP) markers are based on the differential hybridisation of cloned DNA to bulk DNA fragments from restriction-enzyme digestion. Thus, RFLPs are defined by specific enzyme-probe combinations. RFLP markers are codominant. The primary sources of clones for RFLP mapping are cDNA clones and Pstl-derived genomic clones. Genomic clones that represent random sequences may be a poor choice for hybridisation probes because of the large percentage of repeated sequences. RFLPs can detect a virtually unlimited number of markers, thus providing an efficient method for discovering linkages among markers and for constructing genetic maps. This is particularly important in *Prunus* because of the relatively low level of variation typically present in this genus. There are several reports of the use of RFLPs in *Prunus* for map-based selection and for elucidating the extent of genetic variability. However, RFLP analysis has important limitations: it is laborious and time-consuming and it often involves the use of radioisotopes. To overcome some of the difficulties, an alternative called sequence tagged sites (STS) has been developed which is PCR-based but not requiring radioactive probing.

RAPDs

Random amplified polymorphic DNA (RAPD) markers are based on the PCR amplification of random locations in the genome. RAPDs are characterized by using arbitrary primers and permit the quick construction of genetic maps and the saturation of specific genomic regions with molecular markers. A single oligonucleotide is utilized for the amplification of genomic DNA. In contrast to isoenzymes and RFLPs, RAPDs are dominant markers. This feature, as well as their variable degree of repeatability and problems in transferring across populations, limits their utilization primarily to map construction. RAPD techniques have been successfully used in *Prunus* for identifying cultivars, estimating genetic diversity and assessing possible origins for selected genotypes and construction of maps. Problems with DNA quality and a general sensitivity to changes in the reaction conditions can hamper the routine utilization of RAPD markers. These difficulties can be overcome by converting RAPDs to sequence-characterized amplified regions or SCARs. In contrast to RAPD and AFLP methods, SCAR is a PCR-based method that employs specific primers. These primers amplify single bands corresponding to genetically defined loci. SCARs can potentially be converted into codominant markers and are less sensitive to reaction conditions. Different SCAR markers are being evaluated for marker-assisted selection in *Prunus*, including identification of the *Ma1* root-knot nematode resistance gene in Myrobalan plum and the identification of the *Ff* (flesh adhesion) gene in peach.

AFLPs

Amplified restriction fragment length polymorphism (AFLP) technology is a powerful DNA fingerprinting technology based on the selective amplification of a subset of genomic restriction fragments using PCR. DNA is digested with restriction endonucleases and double-stranded specific adapters are ligated to the ends of the DNA fragments to obtain template DNA for subsequent amplification by PCR. The subset of amplified fragments is then analysed by denaturing PAGE to generate the fingerprint. AFLP has a number of advantages over the RAPD technique: more loci analysed per experiment and better reproducibility of banding patters resulting from the higher specificity of primer annealing to complementary adapters. Powell et al. found that AFLPs had a much higher multiplex ratio (number of polymorphic products per "reaction") than other molecular marker systems. Consequently, AFLPs also shows a higher marker index. These markers have been mainly used in *Prunus* for genetic mapping and molecular characterization and estimation of genetic diversity among apricot cultivars.

SSRs

PCR-based, simple sequence repeat (SSR) markers (microsatellites) are becoming the marker of choice for fingerprinting and genetic diversity studies for a wide range of plants. Because of their high polymorphism, abundance, and codominant inheritance, they are well suited for the assessment of genetic variability within crop species, and of the genetic relationships among species. Recently, SSR primers generated from different *Prunus* species have been reported in peach, cherry and almond. These markers are being used for the molecular characterization and estimation of genetic diversity among peach and almond cultivars sweet cherry cultivars and apricot cultivars. SSRs have been employed in the genetic characterization of multiple embryos in almond and genetic mapping. Studies of genetic diversity and genetic relatedness utilizing molecular markers thus offer unprecedented opportunities for improving *Prunus* breeding efficiency when using either established cultivars or interspecific germplasm. Similar opportunities have also been demonstrated for the analysis of chloroplast DNA. SSRs are currently being employed for the molecular characterization and estimation of genetic diversity and the genetic relationships among peach and almond cultivars and related *Prunus* species (Fig. 2). In addition, recent studies have had shown promise for analysing variation of internal transcribed spacers (ITS) in nuclear ribosomal DNA and chloroplast DNA.

Genetic Mapping

Several intraspecific and interspecific *Prunus* maps have been developed using different types of molecular markers. The utilization of PCR-based markers has made mapping and tagging of a wide range of traits possible. The analysis of cosegregation among markers greatly facilitates linkage analysis between markers and major or quantitative loci controlling horticulturally important traits. Different research groups have released linkage maps using isoenzymes, RFLPs and RAPDs in peach, almond, sweet cherry, sour cherry, apricot, and peach × almond hybrids. Similarly, AFLPs allow detection of a higher level of polymorphism in peach than isoenzymes, RFLPs or RAPDs. SSR has also been used for mapping in peach and almond. The first genetic linkage map for a *Prunus* rootstock population was constructed using AFLPs. The similar order of markers observed in different *Prunus* maps suggests a high level of synteny within the genus. This homology among *Prunus* species partly explains the low level of breeding barriers to interspecific gene introgression and supports the opportunity for successful gene transfer between closely related species.

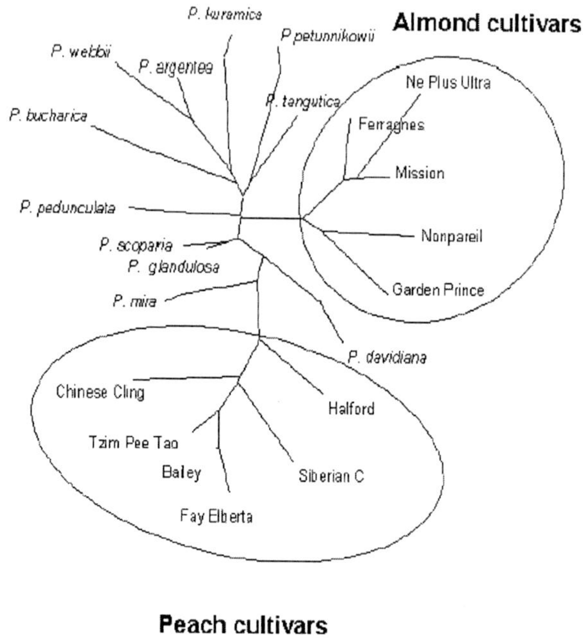

Figure 5. Genetic relationship among peach and almond cultivars and related Prunus species obtained by a study with SSR markers. Unrooted dendogram obtained by Neighbour Joining cluster analysis.

Marker-assisted selection

Marker-assisted selection (MAS) is emerging as a very promising strategy for increasing selection gains. If sufficient mapping information is known, MAS can dramatically shorten the number of generations required to "eliminate" the undesired genes of the donor in backcrossing programs. Marker loci linked to major genes can be used for selection, and sometimes is more efficient than direct selection for the target gene. Selection by molecular markers is particularly useful in fruit and other tree crops with a long juvenile period, and when the expression of the gene is recessive or the evaluation of the character is otherwise difficult, as with resistance to biotic or abiotic stress. The principal approach for the analysis of marker-trait association in *Prunus* is the use of mapping populations segregating for the characters of interest. The different linkage maps developed in *Prunus* (peach, almond, cherry, and apricot) include markers associated with several traits of horticultural value. Mapping quantitative characters by identifying quantitative trait loci (QTL) is also becoming an important tool in tree breeding. QTLs are generally recognized by comparing the degree of covariation for polymorphic molecular marker and phenotypic trait measurements. Important characters and QTLs that are presently being mapped in stone fruits include the control of flower (bloom time, self-incompatibility, pollen-sterility, double-flowers), fruit (shape, pubescence, flesh colour, acidity and sweetness), leaf (red vs. green colour) and tree traits (pillar or weeping architectures), and resistance to various pests and diseases (root-knot nematodes, powdery mildew, leaf curl, Plum pox potyvirus, etc.). The high degree of genome synteny observed among *Prunus* species should also facilitate the successful transfer of sets of markers and coding sequence among species. Bulk segregant analysis (BSA), where two pooled DNA samples are formed from plant sources which have similar genetic backgrounds but differ in one particular trait, is another promising approach for the analysis of molecular marker-horticultural trait association. This method also makes possible the identification of markers linked to the trait of interest. A strategy combining different markers with bulk segregant analysis was used to identify markers linked to loci of specific characters in peach and peach × almond crosses, RAPD markers flanking the red-leaf (Gr) and malate dehydrogenase loci in the NC174RL × Pillar and Marsun × White Glory F2 peach families, and three RAPD markers associated with a delayed bloom gene in almond. Also, it has facilitated the study of self-incompatibility and male sterility in almond. BSA and RAPD analysis were recently utilized to distinguish markers linked to the *Ma1* gene, a major dominant gene that controls a wide-spectrum resistance to root-knot nematodes in Myrobalan plum, as well as markers linked to resistance to PPV in apricot and to ring nematode in peach.

Apart from isoenzymes, RFLP, RAPD, AFLP, and SSR, other markers being used in the development of marker associated traits, are those based on single point mutations (SNPs) and those obtained from either cDNA sequences (ESTs) or databases (Cloned Gene Analogs, CGAs). The first genetic linkage map for a *Prunus* rootstock population was constructed using AFLP technology and, simultaneously, two genes that control resistance to root-knot nematodes, *Mi* and *Mij*, were mapped and tagged. The conversion of the AFLP marker linked to the *Mij* (a gene required for resistance to *Meloidogyne incognita* and *javanica*) locus to STS proved to have practical application for germplasm screening and for breeding peach rootstocks for resistance to root-knot nematodes. An extensive application of molecular marker assisted selection is taking place in the manipulation of self-compatibility in *Prunus*. Most species are predominantly self-incompatibles. Self-incompatibility is of the gametophytic type and acts to prevent self-fertilization. This character is controlled by a single *locus* with multiple codominant alleles, and is expressed within the styles of flowers as *S*-RNases glycoproteins which are responsible of the subsequent inactivation of selfpollen tube growth. Almond self-incompatibility alleles (*S*alleles) were initially identified in the field through controlled crosses with a series of known *S*-genotypes. More recently, molecular methods have been developed in two areas: identification of stylar *S*-RNases by electrophoresis in vertical polyacrylamide gels, and the amplification of specific *S*-alleles using appropriately designed primers for PCR and electrophoresis in horizontal agarose gels. This technique is being routinely used for the identification of crossincompatibility groupings for current almond cultivars and for efficiently breeding self-compatibility into new cultivars (Fig. 6).

REFERENCES

Agarwal, S., and Nath, A.K. 2001. "Characterisation of peach (*Prunus persica* L.) cultivars using isozymes as molecular markers". *Sci. Hort.* 90: 227-242.

Browicz, K., and Zohary, D. 1996. "The genus *Amygdalus* L. (*Rosaceae*): Species relationship, distribution and evolution under domestication". *Genet. Resour. Crop Evol.* 43: 229-247.

Byrne, D.H., and Littleton, T.G. 1989. "Characterization of isozyme variability in apricots". *J. Am. Soc. Hort. Sci.* 114: 674-678.

Gebhardt, K., and Goldbach H. 1988. "Establishment, graft union characteristics and growth of *Prunus* micrograft". *Physiol. Plant.* 72: 153-159.

Gupta, P.K., Balyan, H.S., Sharma, P.C., and Ramesh, B. 1996. Microsatellites in plants: a new class of molecular markers. Curr. Sci. 70: 45-54.

Hesse, C.O. 1971. "Monoploid peaches, *Prunus persica* Batsch: Description and meiotic analysis". *J. Am. Soc. Hort. Sci.* 96: 326-330.

Young, N.D., Zamir, D., Ganal, M.W., and Tanksley, S.D. 1988. "Use of isogenic lines and simultaneous probing to identify DNA markers tightly linked to the Tm-2a gene in tomato". *Genetics* 120: 579-585.

13

Organic Seed Production and Variety Development

A portion of the breeder stock becomes the parent of a larger quantity of foundation stock. The institution associated with the breeder controls the production of foundation stock, and in turn supervises production of registered seed for distribution to licensees, such as seed companies. These companies, in turn, contract (often with farmers) for a large quantity of certified seed. The final stage is production of seed from parent stock of certified (or select) seed for general distribution through commercial channels, although certified seed may be the final stage for large-scale grain production. *Select* is a term used more for vegetable seed, comparable to *certified* for grains. Commercial-scale organic production requires seed stocks (both open-pollinated and hybrid) with proven reliability—especially natural resistance to insects and diseases, as well as natural vigor to germinate promptly and out compete weeds. Good flavor and quality typically are considered more important than shippability. Additional attributes making for successful organic propagation are beginning to be identified.

Recently, organisations such as the Organic Seed Alliance (OSA) and the Public Seed Initiative (Cornell) have outlined a new public participatory model for breeding organic seeds. The model aims to strike a middle course between the inexperience of seed-saving farmers and any special-interest bias in formal research. Prior to training, farmers often lack the skills to select traits important for enhancing organic production. They may also lack resources to carry on multi-year development of seed lines. Leaving the research agenda in the hands of institutions simply accelerates the movement toward genomics and patentable outcomes. In 1999 the Northern Plains Sustainable Agriculture Society (NPSAS) undertook a three-state farmer-driven, participatory breeding program for organic varieties that is still ongoing.

Seeds of Change is leading the way in developing summer squash for organic production, especially zucchinis, emphasising large canopies to shade out weeds, resistance to weather swings, adequate yields, and flavor. A preliminary evaluation of heirloom varieties at Cornell under organic conditions has identified a forgotten cantaloupe with superior flavor. 'Hannah's Choice' thrives under organic conditions, when grown for local markets and not for long-distance shipping.

FARMER COMPENSATION

Exactly how farmers participating in breeding the new organic varieties will be compensated for their time is not clear, except that the farmers will ensure organic versions of their favorite regional varieties for their own use. Neither has anyone offered a clear distribution model for the new varieties. One possibility is the collaborative model (like the California Sweet Potato Growers Group that distributes the virus-free planting material produced by University of California research only to its members). Plant breeding clubs share seeds among their own members, and the membership model has emerged as the preferred method for organic farmers to obtain transplants.

The Organic Seed Alliance calls for "developing new relationships and exploring novel avenues of collaboration to bring quality seed to the organic movement." In the U.S., plant breeding clubs generally include a group of farmers assisted by a university researcher or other technical assistance provider. The Northeast Organic Farming Association of New York (NOFA-NY) continues to work with Cornell University in Cornell's Public Seed Initiative, under a 2004 USDA organic farm research grant, for expansion of on-farm vegetable breeding, on-farm trials, and farmer education to develop and deliver improved vegetable varieties for organic systems.

ISSUES WITH THE CONVENTIONAL SEED INDUSTRY

Heretofore, the increasingly consolidated seed industry has served as the main engine of commercialisation and distribution of new introductions by producing certified (for grain crops) and registered (for vegetable varieties) seed. The industry has sought greater returns for its crucial service by acquiring intellectual property rights to seeds of unique varieties, limiting the number of varieties sold, and most significantly, finding advantageous legal or legislative avenues.

A main attraction of biotechnology for seed companies is enhanced worldwide market share, not improved yields. Accordingly, Gunnar Rundgren, president of the International Federation of Organic Agriculture Movements (IFOAM)—concurring with the assessment of World-Watch Institute—asserts that in the case of GMOs (genetically modified organisms) there are no benefits for either consumers or producers—only for the companies producing and selling them. If farmers feel they need herbicide-resistant varieties, that is because they are locked into a production system that depends on chemical inputs... [a system] that leads to further degradation of the environment, increased dependency of farmers and more risks for everybody.

Acquisition of exclusive ownership of seed varieties is limited under the 1970 Plant Variety Protection Act, which safeguards the rights of farmers and gardeners to use their own saved seed, and the rights of plant breeders to use PVP varieties for breeding new varieties, while affording seed developers a means to recoup their investment. Lobbying groups demanded protection for small farmers in the PVPA legislation.

Seedsaving farmers and gardeners had become concerned by the European ban on many traditional open-pollinated varieties as part of a program of varietal "standardisation." However, under an obscure 2001 U.S. Supreme Court decision (Pioneer Hi-Bred International vs. J.E.M. Ag Supply), companies for the first time could freely patent plant varieties under the 1795 U.S. Utility Patent law, without any reservations to protect small growers or farmers who wished to save (and sometimes sell) seed from their own crops. So far, this has affected mainly U.S. commodity grain crops. At the end of 2004, owners of patents on genetically engineered varieties had fi led 90 lawsuits, involving 147 farmers and 39 small businesses, alleging seed patent violations.

ISSUES IN ORGANIC SEED SOURCING FOR COMMERCIAL GROWERS

In setting as a key goal for the future of public breeding, "development of 'a road map for invigorating public domain plant and animal breeding to meet the needs of a more sustainable agriculture,'" the 2003 Seed Summit committed itself to the totally new area of breeding for organic production. In doing so, it shifted ground beyond increasing the supply of currently available varieties of organic seed to developing new varieties designed specifically for organic production.

MAJOR REGULATORY ISSUES THAT DIRECTLY AFFECT U.S. ORGANIC FARMERS

Seed companies complain bitterly that for the past two years organic farmers have used the availability exemption in the USDA/NOP standards to avoid buying organic seed. Organic seed may be more expensive, and farmers may have to go outside their usual seed sources to find it. Farmers also say that organic seed is simply not available for their preferred varieties. Because the rule that encourages the planting of organic seed is relatively new, many types of organic seed have been in short supply. This situation is improving, as organic production for the seed market grows.

Organic certifying agents differ in their interpretations of this regulation, which simply states that the producer must use organically grown seeds except "when an equivalent organically produced variety is not commercially available." Some certifiers require only that a farmer document three instances in which seed companies that are likely sources for organic seed cannot provide a specific variety. Where a farmer has found organic seed of the desired variety, but it is of poor quality, some certifiers have not required the farmer to use the low-quality seed (i.e., seed with poor germination, low purity, low test weight, etc.). In this instance, the certifier is interpreting the word "equivalent" in the rule to include seed quality characteristics. The quality problem occurs mainly when an organic farmer attempts to use "bin-run," on-farm produced seed that is not certified.

However, in 2005 NOFA-NY began cautioning its certified organic farmers (mainly vegetable growers) to use organic seed. In the fall of 2004 NOFA staff compiled an updated organic seed list that included organic varieties available in 2005 and comparable conventional varieties. For certified organic farmers in the U.S. as a whole, the access problem seems to have been solved for now by the certified organic sourcing service the Carolina Farm Stewardship Association's Save Our Seed Project has begun providing to growers. The American Seed Trade Association (ASTA) has recently met with NOP to request that NOP manage an organic seed database. According to the *Organic Observer*:

ASTA would like to see an interactive database established to provide real-time access to seed suppliers and the public regarding availability of organic seed varieties. ASTA also requested that certifiers be required to supply monthly reports on exemptions granted for non-organic seed. NOP indicated that they are willing to sponsor a database, but are expecting ASTA to provide the data. NOSB members [present] questioned the scope of this project. The problem of varietal "equivalence" has emerged mainly in vegetable production. Seed companies acknowledge that many, practically identical vegetable varieties are sold under different names by different suppliers—in part to get around trademark or copyright issues. Growers have apparently been claiming to their certifiers that an organic variety under a different name is not equivalent to their preferred variety. Other farmers argue that high prices alone exempt them from using organic seed.

GENETICALLY MODIFIED MATERIAL

Requiring testing for GM material is another contentious issue. Some organic grain producers have had export lots rejected by foreign buyers because the lots were contaminated with GMOs. The sheer number of GMOs that have migrated into U.S. food crops leaves the organic industry in a quandary. It's an immediate problem for crops such as canola, soy, and corn, where GMO varieties predominate, and it threatens potential migration of stray GMO material to related weeds and nearby food crops. Two schools of thought have proposed two different solutions. The American Seed Testing Association favors a system of testing organic seed to certify it as GMO-free before it can be planted or sold. On the other hand, the American Seed Trade Association guidelines include this statement:

ASTA strongly supports that organic certification under the NOP is a process, not product certification. . . . ASTA strongly maintains that any movement toward organic seed testing or product certification is not only counter to USDA and NOP policy, but also the U.S. seed industry and organic producers at large. It is well recognised in numerous food and agricultural production standards, including organic standards, that zero is not possible. Furthermore, any movement by seed producers to respond to such unrealistic market demands will not only undermine the viability of the U.S. government's organic policy but could erode the U.S. seed industry's future participation in the organic market.

New procedures are increasingly able to identify GMOs, even in large quantities of seed, with a high degree of accuracy. Some U.S. export grains are tested, and many suppliers of organic grain seed verify that their stocks are free *only to a certain tolerance level*. Tolerances have yet to be set by NOP. Monsanto recently conducted a lab analysis seminar at its St. Louis facility to demonstrate the latest methods of detection. European scientists have detected GMOs in 100% of samples tested. Iowa State University has developed a new software program, using weather data and other geographical parameters, that can predict genetic purity at harvest for hybrid corn in the field, to aid farmers in marketing decisions.

A big problem for on-farm seed producers is that certain crops with GMO analogues already exhibit pervasive, low-level GMO contamination. According to a 2004 study conducted by the Union of Concerned Scientists (UCS) on conventionally produced U.S. soybeans, canola, and corn, representing a wide array of popular varieties with no history of genetic engineering, "more than two-thirds of 36 conventional corn, soy, and canola seed batches contained traces of DNA from genetically engineered crop varieties." The report concluded, "The US may soon find it impossible to guarantee that any portion of its food supply is free of gene-altered elements, a situation that could seriously disrupt the export of US foods, seeds, and oils. Many believe it could also gravely harm the domestic market for organic foods." The lab tests were commissioned by UCS and conducted on certified seed. Many scientists, universities, farmers, and other have questioned plans for GMO wheat. Canola is a major oilseed; domestic corn and soybeans are major ingredients in many products—including starches, emulsifiers, and animal feeds. Some sources have suggested that bacteria can spread GMO material from a genetically engineered crop to a nearby unrelated crop or weed. In fact, this mimics the process used in genetic engineering. These developments raise serious questions about geographically indiscriminate on-farm production of organic seedstocks for grains and oilseeds. Moreover, many varieties of GE crops- including "pharmacrops"—are being grown as trial crops in undisclosed locations in the U.S. As a result, some western organic growers increasingly discriminate among seed suppliers.

INDUSTRY POSITIONS ON TESTING FOR GMOS

Organic spokespeople like Jim Riddle, recently elected to chair the National Organic Standards Board, point out that required testing for GMOs would deeply alter the concept of organics from a process-based system to a testing system. However, there is a marketing issue. The public now believes organic is 100% GMO-free. A system of tolerances for GMO contamination may eventually need to be established for certified organic crops—especially wind-pollinated crops like some grains and oil-seeds. Governmental agreements, especially on harmonisation of organic standards, would open the door for U.S. organic farmers to participate in foreign trade. Other suggestions include setting aside areas of the world still remote enough to produce foundation stock of wind-pollinated crops or establishing a U.S. government public seed bank of pure stock .

QUALITY ISSUES IN FARMER-SAVED AND -TRADED SEED VS. PURCHASED COMMERCIAL SEED

The highest quality grain seed sold to farmers is "certified," with minimum standards for purity, germination, test weight, trueness to type, and absence of physical damage. Ideally, seed for planting organic grain crops would be both "certified" and "certified organic." Shortages of certified organic grain seed have sometimes led farmers to use "bin-run" seed from a nearby organic farm or from a previous year's harvest that (while it is "certified organic") may contain light or broken seed, weed seed and other foreign matter, or pathogens. Such seed is also likely to germinate poorly. This is not invariably the case, of course. According to many certifiers' interpretations of NOP regulations, farmers can by-pass available low-quality organic seed in favor of untreated conventional seed of higher quality.

Value in Going Back to Certified Seed Every Few Years if You Save Your Own

Although Canadian farmer Percy Schmeiser asserts that he selected and saved seed most of his 35 years of growing canola crops—thereby developing a landrace adapted to Saskatchewan conditions—the unfavorable outcome of the internationally publicised court case in which he was involved with Monsanto underscores the advisability of commercial farmers going back every few years to a reliable source of organic seed of their preferred variety. This practice guards against disease buildup, inadvertent contamination of the stock, and reversion of the crop to undesirable traits. This reliable source can be certified seed from a conservator university or commercial seed company. Jeff McCormick, a pioneer new-breed seed company owner, has suggested that vegetable farmers growing a contract seed crop may find it to their advantage to go back to the company every year for "select" (certified) seed for the vegetables they are raising for market, as well.

THE GLOBAL PICTURE

While European Union (EU) and global standards are beyond the scope of this publication, there was extensive discussion of the need for global harmonisation of organic standards at the 2004 World Seed Conference in Rome. Differing standards, of course, affect trade policy, and intense negotiations between the U.S. and the European Union continue.

As of 2005, some GMO plantings in Europe, as well as exports of U.S. Bt corn to Europe, had been approved. Another major issue at the World Seed Conference was intellectual property rights, or the implications of governmentally approved lists of permitted varieties. This is a special concern for traditional farmers in many countries, who are used to saving seed from year to year and have over the centuries developed unique landraces. A recent example is in Iraq, where a new report by GRAIN and Focus on the Global South cites a U.S. edict in occupied Iraq that "prevents farmers from saving their seeds and effectively hands over the seed market to transnational corporations."

Traditional practices of indigenous farmers are mostly compatible with organic production: planting a mix of adapted types (landraces) to ensure some survivors, despite vagaries of weather and insect/disease attacks; use of older varieties geared to minimising capital investment; hand-harvesting and other labor-intensive practices precluded by modern, uniform, machine-harvestable varieties; and use of labor-intensive crop protection strategies like hand weeding and watering, rather than purchased off-farm inputs. But can hand labor feed burgeoning urban populations, or is it a relic of a younger, less densely populated Earth, where 98% of people grew their own food? In the best of all possible worlds, a blend of traits uniquely adapted to organic production (not only resistance to local pests and diseases, but improved vigor and flavor) will result from horizontal breeding. This implies a far more decentralised food production system than we have at present.

Geography of organic seed production has ramifications mainly in the context of GMOs. Spain and Italy raise seed for the rest of Europe. Traditionally U.S. garden seed has been produced in Idaho and other arid West Coast and Intermountain regions. Relative severity of pest and disease pressures is a major consideration in producing quality seed. However, labor costs for seed production became an issue in the 1980s, leading to seed production for commercial growers as far away as Taiwan and Argentina—a development worrisome on several counts, not the least of which is the newly announced Chinese plan to invest billions of dollars in Argentina and Brazil in return for access to land and natural resources, an agreement finalised at the recently concluded Summit in Chile. Argentina has been identified as an emerging leader in GMO crop production.

TUBERS AND ALLIUMS

Commercial growers rarely try to produce their own starts or sets; they rely on specialised suppliers or on grower associations to provide high quality propagation material each year. In 2004 growers temporarily obtained organic vegetable starts from their associations or even from state departments of agriculture, in the absence of commercial production.

REFERENCES

Brook, Rhonda J. 2002. "Pollen tracker" . *Farm Industry News*. mid-February. p. 30–32.

Colley, Michaela, and Matthew Dillon. 2004. "The next great challenge: Breeding seed for organic systems" . *Organic Farming Research Foundation Information Bulletin*. Winter. p. 1, 4, 5, 29.

Guebert, Alan. 2001. "Supreme Court blesses plant patents; bye-bye bin-run seed" . *The Land* (MN). December 21. p. 3.

Jones, Stephen. 2004. "Breeding resistance to special interests" . *OFRF Information Bulletin*. Fall. p. 4–7.

King, Tim. 2004. "Growing organic seed fits farm's rotation" . *The Land*. December 17. p. 9A–11A.

Robinson, R.A. 1996. "Return to Resistance: Breeding Crops To Reduce Pesticide Dependency". *AgAccess,* Davis, California, and IDRC Books, Ottawa, Canada.

Rundgren, Gunnar. 2003. "EU organic seed regulation adapts to realityv. *The Organic Standard*. July. p. 16.

Staff. 2004. "Should there be a GMO tolerance for organic?" *The Non-GMO Source*. April. p. 1–2.

14

Bioinformatics in Plant Breeding

Over the past few decades, major advances in molecular biology and genomic technologies have led to an explosive growth in the biological information generated by the scientific community. This deluge of genomic information has, in turn, generated a need for computerised databases to store, organise and index the data, and for specialised tools to view and analyze the data. With the publication of the complete *Arabidopsis thaliana* genome sequence and the draft sequence for rice genome, plant research and industry have entered the age of genomics. Numerous applications of genomic information have created many opportunities for integrating the rich rewards from sub-systems biology, integrative biology and large scale systematic functional genomics projects. With this accumulation of various types of data, the universe of "genomic understanding" is wide open. With this understanding, it is possible to model and design the amount and sense of changes in the level of gene expression, or how to localise proteins and assess their interactions with other genes and proteins, and finally how they affect the metabolite pools within any given tissue. To reach these goals will require a huge scientific undertaking, many aspects of which will undoubtedly rely on bioinformatics.

Bearing in mind the potential power of data hidden within the complete genome scaffolds, or even within the partial transcriptomics data available for more plant species, it is logical to consider that bioinformatics has become a crucial part of modern genomics research. Bioinformatics is thoroughly involved with the completion and assessment of a multitude of different complete genome sequences. As a science of data management in genomics and proteomics, and as a young discipline in information technology bioinformatics has progressed very fast in the last few years. Bioinformatics is practiced worldwide to access various databases and to exchange information for comparison, confirmation, storage and analysis. To date, there have been several databases on proteins from humans, animals, plants, bacteria, and other life forms.

In biology and medicine, these databases help in developing new inventions which are useful to mankind. Bioinformatics allows life scientists invent new drugs and drug delivery systems, which makes for greater progress in the field of biotechnology. For the future development of biotechnology, bioinformatics will have to take advantage of the internet and the World Wide Web (WWW). Future rDNA research should be guided largely by the databases available for generic or specific forms. Bioinformatics and biotechnology have to move hand in hand to advance. However, bioinformatics can now be considered as a *bona fide* discipline within information technology.

Bioinformatics is the field of science in which biology, computer science, and information technology merge to form a single discipline. The ultimate goal of bioinformatics is to make possible new biological insights and create a global perspective on the unifying principles in biology. At the beginning of Application of bioinformatics in fruit plant breeding the "genomic revolution", the task of bioinformatics was to create and maintain databases to store biological information, such as nucleotide and amino acid sequences. Database development involved not only design issues but the development of complex interfaces whereby researchers could both access existing data and submit new or revised data. Therefore, the most urgent task for bioinformatics today has become the analysis and interpretation of various types of data, including nucleotide and amino acid sequences, protein domains, and protein structures. The actual process of analyzing and interpreting data apples to:

1. the development and implementation of tools which enable efficient access to, and use and management of, various types of information, and
2. the development of new algorithms and statistics with which to assess relationships among members of large data sets, such as methods to locate a gene within a sequence, predict protein structure or function, and cluster protein sequences into families of related sequences.

PLANT BIOINFORMATICS

Plants are the basis of life on earth. They producc the life-supporting oxygen we breathe, they are essential for our nutrition and health, and they provide an environment for the vast biodiversity on earth. For centuries, humans have selected plant varieties that best fit their purposes and developed fruit plants that have many advantages compared to natural, wild plants in terms of quality, quantity and farming practices. However, multifactorial traits involved in resistance and quality have proven to be extremely difficult to improve, certainly in combination. The revolution in life sciences brought on by genomics dramatically increases the scale and scope of our experimental enquiry and applications in fruit plant breeding. The scale and high resolution power of genomics makes possible a broad and detailed genetic understanding of plant performance at multiple levels of aggregation. The complex biological processes that determine pathogen resistance and crop quality are now open for systematic functional analysis. In plant bioinformatics, these analyses are made with the help of special software on huge amounts of data in databases.

ROLE OF MODEL ORGANISM

Over the last century, research on a small number of organisms has played a pivotal role in advancing our understanding of numerous biological processes. This is because many aspects of biology are similar in most or all organisms. It is often much easier to study a particular aspect in one organism than in others. Those organisms which are intensively studied are commonly called model organisms. Each has one or more characteristics that make it suitable for laboratory study. The most popular model organisms have great advantages for experimental research, such as rapid development with short life cycles, small adult size, ready availability, and tractability. They become even more useful when many other scientists work on them. A large amount of information can then be derived from these organisms, providing valuable data for the analysis of gene regulation, genetic diseases, and evolutionary processes in humans and crop plants alike. *Arabidopsis thaliana* is a small flowering plant which belongs to the *Brassica* family, which includes species such as broccoli, cauliflower, cabbage, and radishes. Because *Arabidopsis* has a smaller than other plants and is easily grown in the laboratory, it has become the organism of choice for basic studies on molecular genetics in flowering plants. Scientists expect systematic studies on *Arabidopsis* will facilitate basic research in genetics and molecular biology and will elucidate many questions in plant biology, including some of significant value to agriculture, energy, the environment, and human health.

In the 1980s, there was a growing awareness that investing heavily in studies on different plants such as corn, oilseed rape, and soybean were hampering efforts to fully understand basic properties in all plants. Scientists began to realise that the goal of completely understanding plant physiology and development is so ambitious that it can best be accomplished by turning to a model plant species that many scientists can then study. Fortunately, because all flowering plants are closely related, the complete sequencing of all the genes in a single representative plant species will provide a lot of knowledge about all higher plants. Similarly, discovering the functions of the proteins produced by a model species will provide a lot of information on protein function in all higher plants.

COMPARING GENOME SEQUENCES

The development of techniques for large-scale quantification and identification of biological molecules, together with combined advances in computer technology and the internet have mad available large volumes of biological data that scientists can access from their desktops. By the time the human genome sequence was published in 2001, the rate of DNA sequencing had increased 2,000-fold since the inception of the technology in 1986. The increase was achieved by automation, miniaturisation, and integration of technologies. Applying this approach to other biological molecules including mRNA, proteins, and metabolites has massively accelerated the accumulation of biological data. This data has been made readily accessible, in part due to publications such as the Molecular Biology Database Collection, an annual listing of the best databases publicly available to the biological community. Analysis of the collection reveals a steady growth in the quality and size of the databases. The 2004 edition contains 548 databases classified into eleven categories. A major aim of most genome projects is to determine the DNA sequence either of the genome or of a larger number of transcripts. This both leads to the identification of all or most genes and to the characterisation of various structural features of the genome. Very often a common bioinformatics strategy for sequence alignment is the comparison of cDNA/EST and genomic sequences and annotation. The veracity of any whole genome sequence must be assessed at three levels: completeness, accuracy of the base sequence, and validity of the assembly. In addition to whole genome sequencing, plant sequence data have been accumulating from three major sources: sample sequencing of bacterial artificial chromosomes (BACs), genome survey sequencing (GSS), and sequencing of expressed sequence tags (ESTs).

SEQUENCE ALIGNMENT METHODS AND APPLICATIONS

Sequence alignment is the arrangement of two or more amino acid or nucleotide sequences from one or more organisms so that the sequences sharing common properties are aligned. The degree of relatedness or homology between the sequences is predicted computationally or statistically based on weights assigned to the elements aligned between the sequences. This in turn can serve as an indicator of the genetic relatedness between the organisms.

Sequence Similarity Searching Algorithms: Smith-Waterman is an algorithm for local sequence alignment, using two sequences as input. The difference between NCBI BLAST (also local alignment algorithm) and Smith-Waterman is that (a) BLAST searches for a sequence throughout a database of sequences; and (b) BLAST statistically calculates the most probable match, and Smith-Waterman is calculates the exact match.

Genome Comparison Tools: MegaBlast is an algorithm based on NCBI BLAST for large sequence similarity search. MegaBlast implements a greedy algorithm for the DNA sequence gapped alignment search. MegaBlast is used to compare raw genomic sequences to a database of contaminant sequences, including the UniVec database of vector sequences, the Escherichia coli genome, bacterial insertion sequences, and bacteriophage databases. Any foreign segments are removed from the draftquality sequence or masked in the finished sequence to prevent them from participating in alignments.

Jim Kent's BLAT (BLAST-Like Alignment Tool) is a tool which performs rapid mRNA/DNA and cross-species protein alignments. BLAT is more accurate, 500 times faster than popular existing algorithms for mRNA/DNA alignments, and 50 times faster for protein alignments at sensitivity settings typically used when comparing vertebrate sequences.

Genome based multiple alignment using BLASTZ: BLASTZ is a multiple sequence alignment programme basically used for whole-genome human-mouse alignments. BlastZ output can be viewed with the LAJ interactive alignment viewer, converted to traditional text alignments. LAJ is a tool for viewing and manipulating output from pairwise alignment programmes such as BLASTZ. It can display interactive dotplot, pip, and text representations of the alignments, a diagram showing the locations of exons and repeats, and annotation links to other web sites containing additional information about particular regions.

EST SEQUENCING

ESTs are partial gene sequences which have been or are being generated in several laboratories using different species and cultivars as well as various tissues at different developmental stages. This facilitates the identification of all of the genes expressed in a particular organism, such as the grape and some members of the *Rosaceae* family, including apples (*Malus*), strawberries (*Fragaria*), raspberries and blackberries (*Rubus*), peaches almonds and other stone fruits (*Prunus*). The benefits arising from the rapid generation of large numbers of lowquality cDNA sequences were not universally recognised when the concept was originally proposed in the late 1980s. Proponents of this approach argued that these cDNA sequences would allow for the quick discovery of hundreds or thousands of novel protein coding genes. Critics countered that cDNA sequencing would miss important regulatory elements that could be found only in genomic DNA. The cDNA sequencing advocates appear to have won. Since the original description of 609 Expressed Sequence Tags (ESTs) in 1991, the growth of ESTs in public databases has been dramatic. In mid-1995, the number of ESTs in GenBank surpassed the number of non-EST records in mid-1995. In June 2000, 4.6 million EST records comprised 62% of the sequences in GenBank. Although the original ESTs were of human origin, NCBI's EST database (dbEST) now contains ESTs from over 250 organisms. In addition, several commercial establishments maintain privately funded, in-house collections of ESTs. Throughout the genomics and molecular biology communities, ESTs are now widely used for gene discovery, mapping, polymorphism analysis, expression studies, and gene prediction.

EST sequences are also an important resource for identifying single nucleotide polymorphisms, localising and isolating gene sequences, and for producing cDNA microarrays for expression profile analyses. EST sequencing efforts will be greatly improved by sharing the information held by different laboratories and designing strategies to avoid duplication and extend the coverage of all expressed genes. Expressed sequence tags (ESTs) can be used to discover new genes, map the genome, and identify coding regions in genomic sequences. An EST database consists of ESTs drawn from multiple cDNAs, and there could be potentially many ESTs drawn from each cDNA. In a database like this, ESTs should be partitioned into clusters such that ESTs from each gene are put together in a distinct cluster. A further complication arises because DNA is a double stranded molecule and a gene could be part of either strand. DbEST is a division of GenBank that contains sequence data and other information on "single-pass" cDNA sequences, or Expressed Sequence Tags, from a number of organisms. The Institute for Genomic Research (TIGR) defines TC as Tentative Consensi (assemblies from ESTs) and ET as Expressed Transcripts (both non-human) when building TIGR Gene Indices (TGI).

dbEST release 040805
Number of public entries: 26,605,325

Malus x domestica	183 916
Vitis vinifera	147 300
Prunus armeniaca	15 081
Citrus x paradisi x Poncirus trifoliata	8 002
Vitis hybrid cultivar	6 533
Fragaria x ananassa	5 322
Prunus dulcis	3 864
Citrus reticulata	3 735
Citrus unshiu	2 561
Ananas comosus	1 547
Fragaria vesca	1 306
Citrullus lanatus	693
Citrus clementina x Citrus reticulata	74
Vitis cinerea x Vitis rupestris	61
Cucumis melo	60

Figure 1. Number of ESTs by fruit collected in dbEST (*release 040805*)

TIGR GENE INDICES

The TIGR Gene Indices represent another effort to consolidate EST and other annotated gene sequences. A significant difference between the Gene Indices and UniGene is that the Gene Indices are assemblies of ESTs and other gene sequences rather than clusters. The assemblies tend to represent one transcript, so alternatively spliced products are grouped separately. Furthermore, the process generates a single consensus sequence per assembly. A Gene Index is maintained for fourteen organisms, including man, the mouse, the rat, *Drosophila*, the zebrafish, *Arabidopsis*, and several crop plants, including the grape. Gene Indices are created from publicly available GenBank and dbEST sequences by clustering ESTs with the DNA sequences encoding the coding sequences annotated on DNA and mRNA sequences. ET sequences are extracted from appropriate divisions of GenBank and participate in the clustering and assembly process along with the cleaned ESTs. ESTs and ETs are compared and clustered together if they meet the following criteria: a minimum of forty base pairs match; identity in the overlap region is greater than 94%; and a maximum unmatched overhang of thirty base pairs. These clusters are then assembled into Tentative Consensus (TC) sequences. All sequences that do not belong to an EST cluster are called singletons, and they are used in analysis in rare cases. UniGene is public domain transcriptome database that links ESTs in a cluster if the sequences have a fifty base pair overlap in the 3′ untranslated region (3′ UTR) with 100% identity. These clusters are not run through the more stringent assembly process and consensus sequences are not made.

INTEGRATED WEB RESOURCES FOR FRUIT PLANT GENOMES

Fruit Transcriptome Based Clustering: Taxonomy ID:3750 in NCBI UniGene Database shows the known genes of *Malus x domestica* from GenBank, ESTs from dbEST, and alignments between all transcript sequences. UniGene clustering proceeds in several stages, with each stage adding less reliable data to the results of the preceding stage. This staged clustering affords greater control than a more egalitarian treatment of all links between sequences. There is a range of contemporary genetic marker types and all have been exploited using attributes of EST data. Simple sequence repeats have been identified from the genome data and have applications in genotyping. Single nucleotide polymorphism (SNP) markers have been selected from various EST collections on the basis of available quality scores and, more recently, SNPs have been predicted and validated from various fruits by screening for conserved patterns of polymorphism within EST sequence clusters.

Genome Database for Rosaceae (GDR) is accurate and integrated webbased relational database. GDR contains comprehensive data of the genetically anchored physical map of the peach, an annotated peach EST database, *Rosaceae* maps and markers, and all publicly available *Rosaceae* sequences. Annotations of ESTs include contig assembly, putative function, simple sequence repeats, and anchored position to the peach physical map if applicable. The GDR has been initiated to meet the major deficiency in *Rosaceae* genomics and genetics research, namely a centralised web database and bioinformatics tools for data storage, analysis and exchange. GDR can be accessed at http://www.genome.clemson.edu/gdr/.

MOLECULAR INFORMATION AND FRUIT PLANT BREEDING

Molecular Plant Breeding

As the resolution of genetic maps in the major crops increases, and as the molecular basis for specific traits or physiological responses becomes better elucidated, it will be increasingly possible to associate candidate genes, discovered in model species, with corresponding loci in crop plants. Appropriate relational databases will make it possible to freely associate across genomes with respect to gene sequence, putative function, and genetic map position. Once such tools have been implemented, the distinction between *breeding* and *molecular genetics* will fade away. Breeders will routinely use computer models to formulate predictive hypotheses to create phenotypes of interest from complex allele combinations, and then construct those combinations by scoring large populations for very large numbers of genetic markers. The vast breeding knowledge gathered over the last several decades will become directly linked to basic plant biology, and enhance the ability to elucidate gene function in model organisms. For instance, clearly visible phenotypic traits that are poorly understood at the biochemical level can be associated by high resolution mapping with candidate genes. Orthologous genes in a model species, such as *Arabidopsis* or rice, may not yet be associated with a quantitative trait like that seen in the crop, but might have been implicated in a particular pathway or signaling chain by genetic or biochemical experiments. This kind of cross-genome referencing will lead to a convergence of economically relevant breeding information with basic molecular genetic information. The expected dramatic improvements in phenotypes of commercial interest include both the improvement of factors that traditionally limit agronomic performance (input traits) and the alteration of the amount and kinds of materials that crops produce (output traits). Examples include:

1. abiotic stress tolerance (cold, drought and salt);
2. biotic stress tolerance (fungi, bacteria, viruses, chewing and sucking insects);
3. nutrient use efficiency;
4. manipulation of plant architecture and development (size, organ shape, number, and position, timing of development and senescence);
5. metabolite partitioning (redirecting of carbon flow among existing pathways, or shunting into new pathways).

Rational Plant Improvement

The implications of genomics for food, feed and fiber production can be envisioned on many levels. At the most fundamental level, advances in genomics will greatly accelerate the acquisition of knowledge and that, in turn, will directly affect many aspects of plant improvement. Knowledge of the function of all plant genes, in conjunction with the further development of tools for modifying and interrogating genomes, will lead to the development of a genuine genetic engineering paradigm in which rational changes can be designed and modeled from first principles.

Genotype Building Experiments

Biodiversity determined by the fruit plant genome analysis: In the last few years, an increasing amount of information on DNA polymorphism and sequencing has been accumulated for different plant varieties and cultivars. Most of this information was used for the recognition of different cultivars and for comparing the similarities and differences between them. These distances are measured by the polymorphism on a part of the chromosome whose function is unknown. This type of polymorphism is widely used in genomic studies across the species. The data for the polymorphism are analyzed for a possible link with a quantitative trait of interest of the individual phenotypes. Once such a link is detected, it is called an indirect marker. Indirect markers are closely linked and sometimes overlap with the locus which determines the quantitative trait (QTL). QTLs are defined as genes or regions of chromosomes which affect a trait. QTLs by themselves are difficult to recognise. In both cases, these markers, can be used for further selection. This selection process is called MAS.

QTLs and mapping: The major problem is to define which populations are suitable for QTL-analyses—unstructured and f2 crosses and in plant—large scale populations in order to screen for possible QTLs. As selection is based mostly on markers, a higher mapping density is important. An interval between marker and QTL of about 5 centimorgans (cM) seemed sufficient for effective selection. The simulation studies however showed that selection accuracy dropped down to 81% and 74% with 2 cM and 4 cM distance compared to 1cM. Some advantages of QTL/MAS selection approach come from:

1. measurement of the marker/QTL in early stages of development;
2. low heritability of the trait;
3. for animals—sex limited or measured after slaughtering—meat quality; for plants—malting quality, etc.

How QTL information could be of use?

4. it is assumed that some but not all loci are identified, so selection should be based on the combination of phenotypic and molecular information;
5. in the process of selection the link of markers and traits could decrease so this link should be observed throughout the generations;
6. in the selection process, QTLs prove the simultaneous existence of the desired genes in a line;
7. in crossbred programmes, QTLs could predict the productivity of untested crosses, including their non-additive effect on the information of the parent lines and limited number of crosses;
8. future prospective: with accumulation of molecular data, genotype building programmes will be developed which will set homozygous desirable markers;
9. in introgression programmes for combining the desirable traits from two lines in one;
10. finally, the real world of agriculture is at the stage of accumulation of molecular data.

Analytical approaches: One of the statistical tools for performing the QTL analyses is meta-analysis, which synthesises dense QTL information and refines the QTL position. A programme of this class is the French BioMercator. An environment with complex research opportunities is also PlaNet, the European plant genome database network, which is available at (http://www.eu-plant-genome.net/).

Further development: Further development and detailed discussion on QTLs includes statistical aspects of MAS, setting up the threshold of significance of marker effects, overestimation or bias in estimation of QTL effects, and optimisation of selection programmes for several generations with simultaneous utilisation of MAS and phenotypic data. A specific feature is that detection should be made on specific plant parts such as leaves, roots and fruits, as was proved for grapes.

Efficiency of QTLs

1. Traits of Interest

Experimental results do not always confirm the efficiency of MAS over genotype building. The main reason is the insufficient precision of the initial assessment of a QTL, its location, and its effect. Some QTLs also could be lost in the GB process. For complex productivity traits, the epistatic lost would cause changes in the magnitude of the QTL effect in the parent and progeny generation. It is thus recommended that selection be based on allelic combinations rather on separate QTLs. This is in line with numerous GxE interactions and with selection within the environment of interest for disease or drought resistance. Consequently, efficiency of MAS will depend on the complexity of the species/trait genetic architecture, on the development of the trait in the environment, and on the interactions between them For complex traits, QTLs should be evaluated in different environments. Phenotypic evaluation over consecutive generations is also necessary. Drought resistance seemed to be a more complex trait than disease resistance.

2. Economics

From an economical point of view, the use of markers will be expensive in terms of DNA collection, genotyping, analyses, detection of QTLs, etc. This high price is paid for the genotype building (there is no other way of doing that) and for traits that are expensive to evaluate, such as disease resistance and traits with low heritability.

Species and traits of interest for MAS

Barley: disease resistance, malting quality;

Maize: drought tolerance, earliness, yield;

Rice: disease resistance;

Tomatoes: pest resistance, organoleptic qualities;

Apples (cultivar 'Galaxy'): clones resistant to fungal diseases; W100;

W101

Peaches: results available in the Genome Database for *Rosaceae* (GDR), at http://www.genome.clemson.edu/gdr/

Sustainable Fruit Production and Pomology: à Knowledge-based Approach

Sustainable production is related to obtaining optimum productivity in terms of yield and fruit quality. Two points are of interest:

1. knowledge of the factors which influence productivity;
2. management of the factors to obtain the necessary productivity.

The knowledge is based mostly on accumulation of the data from empirical observations and field experiments, proper planning, and analyses. The aim is to test as many of the factors which might influence important traits. The analyses would reveal the magnitude of different factors and their possible interaction. These factors can be manageable with agrotechnology in all its complexity, or unmanageable because they are random environmental factors, such as weather. Most variation in productivity is caused by these two types of factors. Specific productivity depends mostly on the characteristics of the cultivar. In that sense, the sustainable production is related to the best fit of the manageable elements of agrotechnology to the specific requirements of the cultivars and to the creation of cultivars which are genetically less sensitive to weather conditions. Studies on the sustainable production of orchard species include broad scale experiments, both on-going and already finished. They may be based on techniques such as *in vitro* culture, rooting, and grafting. Development and productivity of individual trees should be observed over their lifetime.

The information to be collected might include all possible observable traits, such as tree development, leaf morphology, branch morphology, growth, flowering, fruit quality, flavor, storability, transportability and resistance to disease and extreme environmental conditions. To complement the studies on DNA polymorphism and sequencing, further QTL analyses are going to be performed on different varieties and cultivars. The observed measurement data should be analyzed for each tree separately. Further analyses will be performed using different schemes in order to reveal possible important influences. These schemes will depend on the traits of interest because some of them cannot be measured individually. The complex of results which is obtained for the influence of different factors on a given trait, on the similarity of influence across the traits, on the link between traits and on the cultivar specificity of these influences are the knowledge base on which proper management of the pomological production system is based. Results of the analyses of this information would give a possibility for qualifying and quantifying the magnitude elements of separate factors. The information obtained for the cultivars of interest could be transferred in a knowledge based system for future planning of the desired productivity. Finally, who needs these results? On one hand, the farmers who are interested in choosing the best cultivars for their specific farming conditions: soil, climate, market conditions, production skills, etc. On the other hand, future selection should be based on information of interest: traits, factors, and influences. Finally, experience in testing could help in designing à proper system for comparing pomological species and cultivars in the future. Proposed collaboration on the knowledge based systems in pomology –specialists on cultivars, molecular biologists and bioinformaticians should collaborate to build a knowledge based system to support decisions in pomological research.

REFERENCES

AGI 2000. "Analysis of genome sequence of the flowering plant *Arabidopsis thaliana*". *Nature* 408: 796-815.

Baxevanis A., Ouellette F. 2001. *Bioinformatics: A practical Guide to the Analysis of Genes and Proteins*. John Willey & Sons, Inc. NY., USA. 518p.

Davenport G., Ellis N., Ambrose M., Dicks J. 2004. "Using bioinformatics to analyse germplasm collections. *Euphytica* 137: 39-54.

Gibson G., Muse S. 2002. *A primer in genome science*. Sinauer Ass. Sunderland, USA, 347p.

Hack C., Kendall G. 2005. "Bioinformatics: current practice and future challenges for life science education". *Biochem. Molecular Biol. Educ.* 33: 82-85.

Morgante M., Salamini F. 2003. "From plant genomics to breeding practice". *Curr. Opin. Biotech.* 14: 214-219.

Rudd S. 2003. "Expressed sequence tags: alternative or complement to whole genome sequences?" *Trenda in Plant Sci.* 7: 321-329.

Sen S., Churchill G. 2001. "A statistical framework for quantitative trait mapping". *Genetics* 159: 371-387.

Smith T.F., Waterman M.S. 1981. "Identification of common molecular subsequences". *J. Molecular Biol.* 147: 195-197.

Walsh B. 2001. "Quantitative genetics in the age of genomics". *Theor. Pop. Biology* 59: 175-184.

15

Breeding of Transgenic Crops

In 1996, approximately 2.8 million hectares were planted to transgenic crops or genetically modified organisms (GMO) in six countries. Adoption has been rapid in those areas where the crops address important production problems, and by 2003 global area had risen to 67.7 million hectares in 18 countries. While this overall rate of diffusion is impressive, it has not been uniform. Developed countries have dominated the use of transgenic crops and only a small number of crops and traits have received commercial acceptance (Table 1). Six countries (the USA, Argentina, Canada, Brazil, China and South Africa), four crops (soybean, cotton, maize and canola) and two traits (herbicide tolerance and insect resistance) account for more than 99 percent of global transgenic area. Sales and licenses in developing countries have been an insignificant source of revenue for industry.

MICROECONOMIC IMPACTS OF TRANSGENIC CROPS

Insect Resistant Cotton

Cotton varieties have been genetically transformed with *Bacillus thuringiensis* (Bt) genes that cause the cotton plant to produce a protein that is toxic to certain species of insects, significantly reducing insect damage and chemical pest control costs in infested areas. Bt cotton is highly effective in controlling two major Lepidopteran pests—pink bollworm and cotton bollworm—and is partially effective in suppressing tobacco budworm and fall armyworm. These four pests are often referred to as the budworm-bollworm complex (BBWC). Bt cotton also offers farmers increased certainty of control because it is effective against insects that have developed resistance to certain chemicals, notably pyrethroids. In certain areas, chemical control of BBWC is no longer effective because of widespread insect resistance to all available chemical pesticides.

Adoption

Transgenic Bt cotton was first grown in the United States and Mexico in 1996 and has subsequently been introduced in Argentina, Australia, South Africa, China, Indonesia and India (Table 2). The first cotton varieties containing a Bt gene were introduced commercially through a licensing agreement between the gene discoverer, Monsanto, and the leading cotton germplasm firm in the United States, Delta and Pine Land Company (D&PL). Some of the same US varieties were subsequently introduced in Argentina, Australia, South Africa, and China.

Table 1: Worldwide adoption of transgenic traits

Global transgenic area crop and trait, 2002	
Product	Share of total transgenic area (percent)
Herbicide tolerant soybean	62
Insect resistant maize	13
Herbicide tolerant canola	5
Herbicide tolerant cotton	4
Insect resistant and herbicide tolerant cotton	4
Herbicide tolerant maize	4
Insect resistant cotton	4
Insect resistant and herbicide maize	4
Total	100

Table 2: Bt cotton area, 2001

Country	Area (000 ha)
United States	2400
China	1500
Australia	125
Mexico	30
Argentina	10
Indonesia	10
South Africa	10
Total	4 085

In the first year of commercial availability in the United States, Bollgardcotton was planted on 850 000 hectares or 15 percent of the total cotton area. By 2001, approximately 2.4 million hectares or 42 percent of US cotton area was planted to Bt or stacked Bt/herbicide tolerant cotton varieties (USDA). However, adoption of Bt cotton has varied greatly across growing regions in the United States, Mexico, and other countries, depending on the availability of suitable varieties and more importantly depending on the particular combination of pest control problems. Bollgardcotton varieties have been rapidly accepted by farmers in areas where BBWC is the primary pest problem, particularly when resistance to chemical pesticides is high. When boll weevils or other pest populations are high, farmers achieve coincidental control of the BBWC with the use of broad-spectrum chemicals, or pesticide mixtures, reducing the value of Bt control. In the United States, adoption has been slowest in California and Texas where suitable Bt varieties have not been available and most rapid in states where chemical pesticide resistance has been most pronounced (Table 3).

Table 3: Percent adoption of Bt cotton in the US, by state 1998-2001

	1998	1999	2000	2001
Alabama	61	76	65	63
Arizona	57	57	56	60
Arkansas	14	21	60	60
California	5	9	6	6
Florida	80	73	75	72
Georgia	47	56	47	43
Louisiana	71	67	81	84
Mississippi	60	66	75	80
Missouri	0	2	5	22
New Mexico	38	32	39	32
N. Carolina	4	45	41	52
Oklahoma	2	51	54	58
S. Carolina	17	85	70	79
Tennessee	7	60	76	85
Texas	7	13	10	13
Virginia	1	17	41	30

Patterns of infestation levels and economic losses also vary widely across the main growing regions in Mexico and have been important determinants of adoption of Bt cotton there (Table 4). While annual infestation levels are variable, Comarca Lagunera and Tamaulipas are the most critically affected by the BBWC and adoption has proceeded most rapidly in these states. The boll weevil is a serious pest in Tamaulipas and South Chihuahua. Pest damage in the other growing regions is more balanced among other insect complexes.

The available evidence indicates that transgenic varieties are scale neutral with regard to both speed of adoption and per hectare benefits. In China, farms of less than one hectare had more than double the net increase in per hectare income of those larger than one hectare (Table 5). This is not surprising given the manner in which Bt cotton and herbicide tolerant cotton and soybean varieties simplify the farmers' management task. In the case of Bt cotton, season-long biological pest control is substituted for chemical control which is temporary. This simplifies the timing and chemical choice decision. The timing and choice of herbicide is similarly simplified for herbicide tolerant crops. Glyphosate effectively controls both broadleaf weeds and grasses and has a fairly broad window for the timing of application.

Table 4: Adoption of Bt cotton and geographic distribution of pest problems in Mexico's major cotton areas.

Pest	Bt effectiveness	Alternate plant hosts	Seriousness of problem					
			Comarca Lagunera	Tamaulipas	North Chihuahua	South Chihuahua	Sonora	Baja Calif
Pink bollworm	100%	none	Highest	none	minor	medium	medium	medium
Cotton bollworm	High	maize, tomato	High	high	medium	medium	minor	minor
Tobacco budworm	Partial	maize, tomato	Medium	high	medium	medium	medium	minor
Army worm	Partial	many	Minor	high	medium	medium	minor	minor
Boll weevil	None	none	Eradicated	highest	minor	highest	minor	none
White fly	None	many	Minor	none	none	none	highest	highest
2000 Bt adoption	–	–	96%	37%	38%	33%	6%	1%

Source: Traxler, et al.

Table 5: Distribution of benefits of Bt cotton adoption by size of farm or income class in China, 1999

	Bt as % of Observations	Yield Increase kg/ha	Change Total Cost RMB/ha	Change in Net Income RMB/ha
Farm Size				
0.0-0.47 ha	86	410	–1346	3331
0.47-1 ha	85	–134	–4429	3871
1+ ha	87	–124	–1510	1534
Household income (RMB)				
1–10,000	85	170	–2503	3151
10,000+	91	65	–449	1301
Per Capita Income (RMB)				
1–1,500	85	456	–1784	3702
1,500–3,000	83	8	–2355	2519
3,000+	97	–60	6	–125

The official exchange rate between the renminbi (RMB) and U.S. dollar is $1.00 = RMB 8.3

Source: Pray and Huang (2003)

Farm and Aggregate Economic Impacts of Bt Cotton

There is great annual and geographic fluctuation in estimates of the actual yield performance difference between Bt and conventional cotton. Insect infestations vary widely across time and space, and the relative performance of Bt cotton is highest when pest pressure is heaviest. Several studies contain extensive reviews of the completed research which we summarise here. An important advantage of Bt over chemical control of pests is that it prevents insects from damaging the cotton plants because the Bt control is always present in the plant. Bt crops also benefit from decreased dependence on weather conditions that affecting the timing and effectiveness of insecticide applications. Chemical pest control is less effective and will generally result in crop losses in the presence of insect infestation. Farmers apply chemical controls only after noticing the presence of pests on the cotton plants, by which time some damage has already occurred. As a result, Bt varieties have superior yield performance over a wide range of growing conditions.

Field level studies of the performance of Bt cotton have also been completed in Mexico, Argentina, Australia, South Africa, China, and India. The results are consistent with the findings of the studies that have been done in the United States (Table 6). In five of the six countries, Bt cotton varieties had higher yields, were more profitable and saved on pesticide expenditures. Qaim and Zilberman argue that the relative performance of Bt cotton is likely to be highest when used by developing country small farmers because of the large pest losses suffered by these farmers. This notion is supported by the international data available to date showing the yield advantage to be largest in India, China and Argentina.

Table 6: Performance differences between Bt and conventional cotton, 6 countries.

	Argentina	Australia	China	India	Mexico	S. Africa
			(Bt – conventional)			
Lint Yield (kg/ha)	+531	–122	+523	+699	+170	+91
Chemical sprays (No.)	–2.4	–4.8	n/a	–3.0	–2.2	n/a
Revenue ($/ha)	+$ 121	n/a	+ $ 13	n/a	+$ 248	n/a
Pest control costs ($/ha)	–$ 34.85	–$ 80	–$ 230	–$ 30	–$ 157	+$ 4.5
Profit/hectare ($/ha)	$ 22.89	–$ 46	+ $ 470	n/a	+$ 335	+$ 29

n/a: information not available

Sources: Qaim and deJanvry, 2003; Fitt, 2001; Traxler et al., 2003; Ismael et al., 2001; Pray et al., 2002; Qaim and Zilberman, 2003.

Several studies have estimated the aggregate impact and the functional distribution of benefits from the introduction of transgenic varieties on benefits to producers, consumers and industry. These studies use estimates of the farm level cost savings and model world cotton supply and demand within an economic framework to calculate benefits. This framework takes account of the fact that, as the new technology reduces the cost of production, farmers may expand supply and that as prices drop, consumers may demand slightly more cotton. These price changes affect the level of calculated benefits. Part of the motivation for these studies has been that, except for a few varieties in China, the

Bt cotton transgenics have all been patented private sector innovations. Patent holders may hold some monopoly power over pricing of their innovation. Certainly, the price of transgenic seed has been higher than that for seed of conventional varieties, and technology fees are charged on top of such high prices for GM seeds. Does this mean that the marketing firms are extracting all of the benefits generated by the innovation? This is an unlikely outcome because farmers must be receiving some benefits, or they would not choose to adopt. It will generally be true that an innovator will only be able to extract part of the economic benefits created through their research effort. There will always be benefit "spillovers" to be enjoyed by other members of society. The empirical studies that have been completed find that the benefits from biotechnology innovations have been widely shared among consumers, producers and industry.

Studies by Falck-Zepeda et al. and Traxler et al. calculate the annual distribution of benefits among cotton producers, consumers and germplasm suppliers for the 1996-98 period using a standard economic surplus model. The estimated amount and distribution of benefits from the introduction of Bt cotton fluctuates from year to year, but total annual benefits created averaged approximately $215 million (Figure 1). The average benefit shares were 45 percent to US farmers, 36 percent to germplasm suppliers and 19 percent to cotton consumers. Frisvold et al. use a different modeling approach to calculate aggregate welfare changes from the introduction of Bt cotton in the same period. They estimate a smaller amount of average total benefits ($181 million), and a smaller share of benefits to US farmers (20 percent) and more to US consumers (27 percent). The share of benefits to industry is estimated at 38 percent.

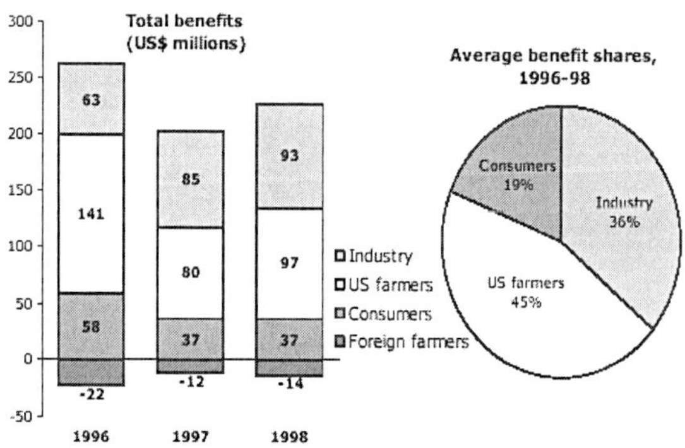

Source: Falck-Zepada, Traxler & Nelson

Figure 1: Benefit distribution from Bt Cotton in US ($ millions)

The average benefit shares from the introduction of Bt cotton in the Comarca Lagunera region of Mexicowere 16 percent for germplasm suppliers and 84 percent for farmers (Table 7). The per hectares change in variable profit accruing to farmers varied widely between the two years, with an average figure of $ 335.45. Therefore, for the two years, we estimate that a total of more than $ 6 million in benefits was produced. In this calculation as in the welfare calculations for the United States, not the entire amount attributed to Monsanto is truly a net benefit, because costs such as seed distributor compensation, administrative and marketing costs were not accounted for. The $1.5 million revenue from seed sales is not a large sum for a company such as Monsanto with $5.49 billion in annual revenue. The large annual fluctuations are largely caused by variability in pest infestation levels – in years of heavy pest pressure, Bt cotton produces a large advantage over conventional cotton varieties. Because Mexico grows a small share of the world's cotton, there is no effect on consumers' benefits.

Table 7: Estimates of economic benefit distribution, Comarca Lagunera region of Mexico, 1997 and 1998 ($US)

		1997	1998	Average
A	Conventional seed price per kg	2.21	2.21	2.21
B	Cost per hectare to produce Bt seed	30.94	30.94	30.94
C	Monsanto/D&PL Bt revenue per hectare	101.03	101.03	101.03
D=C-B	Monsanto/D&PL net revenue per hectarea	70.09	70.09	70.09
E	Change in farm variable profit per hectare	44.15	626.74	335.40
F	Bt area in Comarca Lagunera	4,500	8,000	6,250
G=D*F	Monsanto/D&PL total net revenuea	315,420	560,747	438,083
H=E*F	Total farmer benefits	198,677	5,013,952	2,606,314
I=G+H	Total benefitsa produced	514,097	5,574,699	3,044,398
J=G/I	Monsanto/D&PL share of total benefitsa	61%	10%	14%
K=H/I	Producer share of total benefits	39%	90%	86%

Effect of Bt Cotton on Use of Chemical Pesticides

As noted above, Bt cotton is totally or highly effective in controlling several lepidoptera species known as the budworm-bollworm complex (BBWC)–the pink bollworm (Pectinophora gossypiella), cotton bollworm (Helicoverpa zea)–and is partially effective in controlling tobacco budworm (Heliothis virescens) and fall armyworm (Spodoptera frugiperda). In many major cotton-growing areas, BBWC is a major or the major pest control problem, but pesticide use is also conditioned by the presence of other cotton pests such at boll weevil. As a result, the effect of the introduction of Bt cotton on pesticide usage varies from region to region. In areas where BBWC is a major pest Bt varieties have contributed to a dramatic reduction in pesticide use. In the United States, the number of pesticide applications used against BBWC has fallen from 4.6 in 1992-95 to 0.8 applications in 1999-2001 (Figure 2).Carpenter and Ginanessi estimate that the average annual reduction in use of pesticides on cotton in the United States has been approximately 1,000 tons of active ingredient. Pray, et al. estimate that in 1999-2001 farmers applied an average of 43.8 kg/hectares less of formulated product on Bt cotton China. Pesticide use also declined in Mexico as Bt cotton use grew from 0 in 1995 to 96 percent in 2000 (table 8). Qaim and Zilberman found that less toxic pesticides, as well as a smaller total quantity of pesticides were used in farmer-managed field trials of Bt cotton in India in 2001. Three less applications and nearly 70 percent less total insecticide was used on Bt cotton, with most of the reduction occurring in highly toxic organophosphates, carbamates and pyrethroids belonging to toxicity classes I and II. Less pesticide use on Bt cotton was also reported in Argentina (Qaim and deJanvry) and South Africa (Kirsten and Grouse).

Table 8: Average number of insecticide applications targeted to principal cotton pests in the Comarca Lagunera, 1995-2000

Year	Pink Bollworm	Tobacco Budworm	Conchuela	Fall Army worm	White Fly	Total
1995	3	2	0	1	1	6
1996	7	2	0.3	2	2	7.35
1997	1.5	2.5	2	1.5	0.4	5.1
1998	2.5	1.3	1	2.1	0.2	4.5
1999	0	0	2	1	1	3.5
2000	0	1	1.5	0.2	0	2

Source: Sánchez-Arellano, 2000. Data from Plant Health Authority insecticide use records.

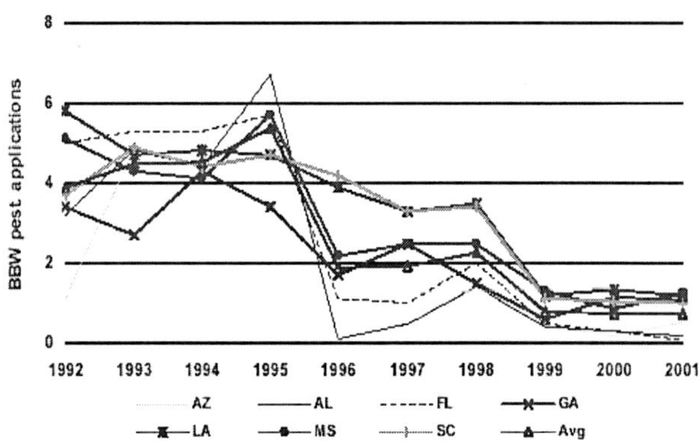

Figure 2: The number of pesticide applications for budworm-bollworm complex, selected US states, 1992-2001

Herbicide Tolerant Soybeans

Genetically engineered herbicide tolerant crops feature a gene from the soil bacterium *Agrobacterium tumefaciens,* which makes the recipient plant tolerant to the broadspectrum herbicide glyphosate. Introduced to a crop plant, the technology can facilitate weed management in farmers' fields. It can reduce production costs, through the substitution of glyphosate for an array of more expensive herbicides. Herbicide tolerance for various crops was developed by Monsanto under the brand name RoundupReady (RR). Monsanto also sells glyphosate under the brand name Roundup. Roundup sales increased significantly after the launch of the technology, however Monsanto's patent on glyphosate has expired and a number of firms now offer the product under various brand names.

Adoption

RR soybeans were commercially released in the Argentina and the United States in 1996. The sale and use of RR technology is protected in the US through patents and sales contract with farmers, but neither form of intellectual property protection is used in Argentina. Thus in Argentina, RR soybeans are widely available from sources other than Monsanto and Argentine farmers pay a relatively small price markup. Argentine farmers are legally allowed to use farm-saved seeds. Adoption proceeded rapidly in both countries. By 2001, more than 90 percent of the Argentine soybean area, and 68 percent of US area was cultivated with RR seeds.

The market for RR certified soybean seeds in Argentina can be characterised as competitive due to an unusual set of circumstances surrounding the introduction of the RR gene into Argentina. The first company to commercially release RR soybean varieties in Argentina was Nidera, the largest seed company in Argentina. Because Monsanto failed to patent the RR technology in Argentina, Nidera obtained royalty-free access to Monsanto's RR technology in the late 1980s (Qaim and Traxler). Nidera channeled the technology through the Argentine biosafety process and received commercial approval for several RR soybean varieties in 1996. Monsanto itself and other companies only followed in subsequent years. In 2001, there were seven companies providing over 50 different RR varieties in Argentina. Except for Nidera, these companies pay license fees to Monsanto. Thus, both Nidera and Monsanto capture some revenue from RR technology.

Farm and Aggregate Economic Impacts of Herbicide Tolerant Soybeans

Argentine farmers are not required to sign special purchase contracts, as used by Monsanto in the United States. This means that farmers are allowed to retain seeds from their harvest for future plantings. The national seed institute in Argentina, INASE, estimates that farm-saved seeds account for 30 percent of all soybeans planted. Although sales of farm-saved and other uncertified materials are prohibited under national law, unauthorised sales are estimated to account for another 35 percent of total seed consumption. The remaining 35 percent are certified seeds sold by authorised seed companies. Weak intellectual property protection and the widespread use of farm-saved and black market seeds have placed downward price pressure in formal seed markets in Argentina. While farmers in the United States on average pay 43 percent more for RR than for conventional certified seeds, the technology markup is around 30 percent in Argentina.

Yields of RR soybeans are not significantly different from yields of conventional soybeans in either the United States or Argentina. Reduced herbicide and tillage expenses generate farm level benefits of RR soybeans. Many farmers switched to low-till or even no-till cultivation practices after adoption of RR soybeans and machinery and labor costs are also lower due to the reduced time needed for harvesting. Due to the lower incidence of green weeds in RR plots, the combine harvester can be operated at higher speed without the danger of clogging.

In Argentina total variable cost of production is about eight percent ($21 per hectare) lower for RR soybeans than for a conventional crop. In the United States for 1996, Hubbell et al. reported cost savings between $17 and $30 per hectare for the US as a whole. Moschini et al. estimated a cost advantage of $20 per hectare for 2000. Duffy carried out farm surveys in Iowa in 1998 and 2000, and found that cost savings are actually negligible there. But, by simulating different weed control scenarios, Gianessi et al., calculated RR cost advantages of $40 per hectare for many US states, in some cases even higher than this. The different results do not suggest a clear pattern over time. Taking an average over all sources, it appears that cost savings in the United States are similar to those in Argentina, even though the prices for RR seeds and glyphosate are lower in Argentina than in the United States.

Welfare effects of the spread of RR soybeans in the United States have been analyzed in a few studies but only Qaim and Traxler has explicitly modeled the diffusion of the technology in Argentina and results from this study will be summarised here. In 2001, RR soybeans created more than $1.2 billion, or about 4 percent of the value of the world soybean crop, in economic benefits at the global level (Table 9). The largest share of these overall benefits went to soybean consumers, who gained $652 million (53 percent of total benefits) due to lower prices.

Soybean producers received net benefits of $158 million (13 percent), and biotechnology and seed firms received $421 million (34 percent) as technology revenue. Soybean producers in countries where RR technology is not available faced losses of $291 million in 2001 due to the induced decline of about 2 percent ($4.06/ mt) in world market prices. This underlines that national restrictions to GM technology access can bring about considerable taxation of the domestic farm sector. A case in point is Brazil, the second largest soybean producer in the world. According to industry estimates, farm level benefits in Brazil could be similar to those in Argentina. Yet, due to a protracted biosafety process and uncertainty with respect to legal responsibilities, as of early 2003, RR soybeans have not received official approval for commercialisation in the country.

Table 9: Benefits generated from the introduction of RR soybeans (million US$)

Year	Benefits to producers	Benefits to Consumers	Technology revenue	Total benefits and technology revenue
Argentina				
1996	–1.2	0.1	0.0	–1.1
1997	9.6	0.8	1.4	11.8
1998	64.5	2.1	10.5	77.1
1999	144.5	3.0	18.4	165.8
2000	201.2	4.7	23.3	229.2
2001	303.2	4.3	27.6	335.0
United States				
1996	10.7	3.2	9.9	23.7
1997	70.5	24.4	70.0	165.0
1998	166.2	67.1	208.7	442.0
1999	136.5	99.0	271.1	506.7
2000	113.9	117.3	304.9	536.1
2001	144.9	149.4	393.1	687.4
Rest of the world				
1996	–5.2	9.1	—	4.0
1997	–46.5	75.2	—	28.7
1998	–138.2	228.8	—	90.5
1999	–187.3	354.2	—	166.9
2000	–215.0	387.8	—	172.8
2001	–290.6	498.0	—	207.4
Total				
1996	4.3	12.4	9.9	26.5
1997	33.6	100.4	71.5	205.5
1998	92.4	297.9	219.2	609.6
1999	93.7	456.2	289.5	839.4
2000	100.2	509.9	328.2	938.2
2001	157.5	651.7	420.7	1,229.8

Source: Qaim and Traxler, 2003.

Farmers in Argentina and the United States had large welfare gains that increased as RR adoption increased. Argentine farmers were receiving surplus of more than $300 million by 2001 and US farmers received surplus of $145 million in 2001. Although the RR area in the United States is larger than in Argentina, net producer surplus has been larger in Argentina since 1999 because the share of adopting farmers in Argentina exceeds the share in the United States. For example, in 2001, more than 9.6 million hectares were still planted to conventional soybeans in the United States, compared to only about 1 million hectares in Argentina. This example clearly shows that, because of technology spillovers, producer benefits are not confined to the innovating country.

Farmers in developing countries have much to gain when they are given access to suitable foreign technologies. Table 9 also shows that monopoly rents for private firms in the United States are sizable, and that their share in total national surplus has been increasing from 42 percent in 1996 to 57 percent in 2001. On the other hand, because of weak intellectual property protection in Argentina, technology revenues there are much smaller, accounting for just 8 percent of the total Argentine surplus. Falling prices for RR seeds and a growing informal market will further reduce this revenue stream over time. However, these results also show that the notion that private firms would gain nothing from their innovations without patent protection is not accurate. Given the big market size, Argentina will remain interesting for foreign seed companies, even though intellectual property protection is weaker than in the United States.

RR Soybeans: Environmental Effects

RR soybeans change the use patterns of tillage and chemical herbicide use. Glyphosate substitutes for a number of other products, with the result that per hectare herbicide expenditures decline. Table 10 shows that in Argentina the average number of herbicide applications slightly increases, while herbicide amounts used per hectare even more than double. In the United States, the use of RR soybeans has been reported to lead to a decrease in the number of applications, with aggregate herbicide amounts more or less unaffected. Herbicides differ in their mode of action, duration of residual activity, and toxicity, so an increase in total herbicide amounts does not inevitably entail negative environmental effects. Glyphosate essentially has no residual activity and is rapidly decomposed to organic components by microorganisms in the soil. According to the international classification of pesticides, glyphosate belongs to toxicity class IV, the lowest class for "practically non-toxic" pesticides.

As Table 10 shows, adoption of RR soybeans led to an almost complete abandonment of herbicides belonging to toxicity classes II and III. There are no other herbicides used in soybeans which belong to toxicity class I. Consequently, RR technology has led to an increase in the use of a relatively harmless herbicide and a significant reduction in the use of more hazardous herbicides. The major reason for the rise in the number of herbicide applications is the farmers' conversion to no-till practices that require pre-seeding chemical weed control. Although RR soybeans were not the only factor for the rapid adoption of no-till practices in the second half of the 1990s, Table 10 suggests that they played an important role. Whereas only 42 percent of the farmers in our sample used no-till for conventional soybeans, 80 percent of them use this practice on their RR plots. No-till helps to preserve the soil texture and reduces the risk of wind and water erosion, with concomitant positive environmental effects. RR farmers who did not completely switch to no-till usually pursue a reduced-tillage system for soybeans. On average, the technology reduced the number of tillage operations by one passage per field. Overall, the number of machinery hours is reduced by 20 percent, and fuel savings are almost 10 liters per hectare.

Table 10: Herbicide use effects of RR soybean adoption in Argentina

	Conventional soybeans (n = 59)	RR soybeans (n = 59)	Percent change
Number of herbicide applications	1.97	2.30	16.8
Total amount of herbicides (l/ha)	2.68	5.57	107.8
Herbicides in toxicity class II (l/ha)	0.42	0.07	–83.3
Herbicides in toxicity class III (l/ha)	0.68	0.00	–100.0
Herbicides in toxicity class IV (l/ha)	1.58	5.50	248.1
Share of farmers using no-till practices	0.42	0.80	90.5
Number of tillage passes per plot	1.66	0.69	–58.4
Machinery time (h/ha)	2.52	2.02	–19.8
Fuel (l/ha)	53.03	43.70	–17.6

Source: Qaim and Traxler, 2003

INSTITUTIONAL OPTIONS FOR DELIVERING BIOTECHNOLOGY INNOVATIONS

The impact of transgenic technologies on production costs will influence comparative advantage in world markets. The direction and causes of shifting trade patterns are difficult to determine due to the many random influences and large annual trade flow variations. However, it is clear that unequal access to the fruits of biotechnology research will lead to differential rates of technological progress between "have" and "have not" countries. To date, the realisation of production cost savings from biotechnology research have been concentrated in just three large countries that are exporters of major food crops: the United States, Canada and Argentina. The primary explanation for this geographic concentration is the dominance of the private sector in the delivery of biotechnology innovations to date, the huge size of these seed markets and the similarity of these countries' agricultural systems. The private sector naturally has focused their R&D effort on those markets with the greatest sales potential for seeds.

The private sector led progress in biotechnology has been financed by corporate debt and equity, commercial sales and through specialised "start up" firms using venture capital. The public sector has played a role in supporting basic research. The total US federal and state public sector budget for ARS and the Land Grant Universities for agricultural research was more than two billion dollars. Developing country potential to finance biotechnology investments occurs on a vastly smaller scale. Incentives are weak for private sector investment even in large developing countries such as China and India because of uncertainty about the potential to enforce IPR. Public sector research efforts are also small relative to the US. James estimates that 96 percent of the $4,400 million invested globally in crop biotechnology in 2001occured in industrialised countries, and that the private sector invested just $36 million in developing countries compared to $3,100 million in industrial countries (Table 11).

Table 11: Estimates of global R&D expenditures on crop biotechnology, 2001 ($ millions)

Industrial	4,220
Private	3,100
Public	1,120
Developing Countries	180
China	115
India	25
Brazil	15
Others	25
Total	4,400

Source: James, 2002a

By comparison the CGIAR, which is the largest international public sector source of agricultural technologies, spends less than 250 million dollars annually on all plant improvement R&D and only about $25 million on biotechnology. Only in China has the public sector been a significant source of innovation in biotechnology. To date, no GM crops have been introduced into tropical regions. The geographic and crop focus of private sector biotechnology R&D and innovation is unlikely to change soon, and the rate of innovation in countries that do not have large seed markets will undoubtedly continue to lag.

Each Country Produces its Own GMOs

This is the least likely of the institutional options. The capacity needed to use technology different from capacity needed to generate technology. Even large countries such Argentina, South Africa and Mexico that are using GMOs, lack the capacity to generate a stream of GMOs targeted at their individual domestic agricultural problems. The GMOs that they are using were developed by Monsanto for the United States market, so the only research needed was adaptive plant breeding to transfer the Bt and herbicide tolerance genes to local commercial varieties. Certainly there are very many small countries that lack the capacity for even this type of adaptive research.

CGIAR Centers Lead Delivery

This is an intriguing institutional possibility given the success of the CGIAR centers in distributing the semi-dwarf wheat and rice technologies that induced the Green Revolution. But the Green Revolution and the Biotechnology Revolution are arriving in vastly different technologies and are arriving in vastly different worlds. The dwarfing genes were freely shared among scientists in many countries without regard to intellectual property rights, were widely adapted and easily moved to adapted germplasm at low cost and faced few phytosanitary concerns. The CGIAR centers are not heavily invested in biotechnology research capacity at present with a total annual expenditure of perhaps $25 million. It therefore seems unlikely to expect them to be a major supplier of biotechnology research for developing countries.

Regional NARS Lead Delivery

Another institutional possibility would be for the larger National Agricultural Research Systems (NARS)–India, China and Brazil–to become regional suppliers of biotechnology research for smaller countries. The advantages of clustering research efforts for countries with similar agroclimatic conditions are obvious, and each of these countries has significant research capacity in both basic and agricultural science. But of these countries, only China has experience with public sector delivery of a transgenic product, in fact Brazil and India have only recently approved GMOs for commercial use. There is no indication that the public sector in any other country will soon become a major player, and no other country has yet benefited from biotechnology discoveries made in China.

The lack of institutional arrangements for sharing intellectual property across public sector institutions is also a large hurdle to be overcome. Contrary to the pace at which private sector companies now share intellectual property, there is scant experience anywhere in the world where public sector institutions have the flexibility or the motivation to achieve such exchanges. This implies that a radically new mindset and new institutional arrangements would need to emerge before the sharing of intellectual property could become sufficiently routine to allow smaller countries to depend on their large public sector neighbors to supply useful research outputs. At present, outside of germplasm being shared within the CGIAR networks, there is very little cross-border sharing of technology between public sector institutions, probably due to a lack of incentives for public officials to negotiate such arrangements, but also probably due to the implicit competition among countries in international commodity markets.

Private Sector Delivery

This is the tested model, having delivered current GMOs. At least three major obstacles may prevent this from becoming a viable option in the near term. First, transaction costs are very large for market entry in each country. In most countries obtaining biosafety clearances is either impossible or so uncertain and expensive that the private sector does not consider market entry to be a good business risk. The list of countries with functioning biosafety committees is increasing, but until there is some type of regional harmonization and sharing of biosafety information, the regulatory transaction costs are entry barriers for a substantial number of countries. A second obstacle is the difficulty of protecting IPR. The experiece to date with IPR enforcement on soybean, maize and cotton GMOs worldwide is mixed—protection has been good in some countries, difficult in others, and uncertain in most. The third, and possibly the most difficult to remedy obstacle is the absence of functioning seed markets in most countries for most crops. With the exception of maize, cotton and vegetables in a few countries, seed markets are very thin, making it difficult to deliver GMOs to farmers. The combined effect of these obstacles is an environment of very weak incentives for private sector biotechnology investment in developing countries—certainly nothing like the business potential that fueled the discovery of existing GMOs. At present then, the developing world suffers from the absence of profitable GMO markets with secure access.

The one scenario under which the private sector could become a reliable source of biotechnology innovations for developing countries is the one in which the large market countries, i.e. India, Brazil and China become "GMO friendly." That is, if India, China and Brazil were to achieve stable regulatory and IPR regimes and GMO products were accepted by consumers in these large markets, the private sector would very likely make substantial R&D investments in developing GMO products for their significant agricultural problems. These three countries have a combined seed market of $4.7 billion. Products developed for these markets would then become available for neighboring countries that have mounted the necessary biosafety regulatory and IPR enforcement capacity. Once the private sector had developed useful products, for example for Brazilian farmers, they would begin marketing them in other tropical countries. Even if this scenario were to develop, the prospects for pureline crops would remain very uncertain.

REFERENCES

Alston, J.M., Norton, G.W., & Pardey, P.G. (1995). *Science Under Scarcity: Principles and Practice for Agricultural Research Evaluation and Priority Setting,* Ithaca NY: Cornell University Press.

Byerlee, D. and K. Fischer. 2001. "Accessing Modern Science: Policy and Institutional Options for Agricultural Biotechnology in Developing Countries," *IP Strategy Today,* No. 1-2001.

Carpenter, J.E. and L.P. Gianessi. "Agricultural Biotechnology: Updated Benefits Estimates, National Center for Food and Agricultural Policy: Washington, D.C., January 2001.

Hubbell, B. J., Marra M. C., and Carlson G. A., 2000. "Estimating the Demand for a New Technology: Bt Cotton and Insecticide Policies. *Am. J. Agric. Econ.* 82, 118-132.

Ismael, Y., Bennett, R. and Morse, S.. (2001), "Farm level impact of Bt cotton in South Africa." *Biotechnology and Development Monitor,* No. 48, p. 15-19.

James, C. 2002a. "Preview: Global Status of Commercialised Transgenic Crops: 2002", ISAAA Briefs No. 27. ISAAA: Ithaca, NY.

Kirsten, J. and M. Grouse. 2003. "The Adoption and Impact of Agricultural Biotechnology n South Africa." in *Economic and Environmental Impacts of First Generation Biotechnologies* ed. Nicholas Kalaitzandonakes, Kluwer Academic Press/Plenum Publications.

Marra, M.C., P.G. Pardey, and J.M. Alston, eds. *The Payoffs to Agricultural Biotechnology: An Assessment of the Evidence.* Washington: International Food Policy Research Institute, 2002.

Pray, C.E., D. Ma, J. Huang and F. Qiao. "Impact of Bt Cotton in China." *World Development,* 29:5 (May 2001).

Shelton, A.M., J.Z. Zhao, and R.T. Roush. "Economic, Ecological, Food Safety, and Social Consequences of the Deployment of Bt Transgenic Plants." *Annual Review of Entomology,* 47:845-81, 2002.

16

Future of Crop Breeding Technologies

Plant breeding started as purely an art (i.e., experience and good sense). It moved on to become a science-based technology that used statistical theory for field plot testing, quantitative genetics to plant breeding operations, and that made some use of single factor genetics. It has continued to evolve, and now is highly mechanised and computerised. Hand labor is nearly eliminated for many operations, and large portions of the data now are gathered computer-ready in the combine, or in robotised and computer-run laboratory operations. Most recently, biotechnology has become an essential tool for insertion of novel genes, for tracking germplasm movement in backcrossing programmes, and for precise identification of genetic material that is protected with intellectual property rights.

None of the early technologies have been dropped, all remain necessary. Therefore, a well-rounded plant breeder today must know many more kinds of science and technology than were required as recently as 10 years ago, not to mention 20 or 30 years earlier. In actuality, no one person can master all of the science and technology that is needed for efficient plant breeding for the late 1990s. At best. a well-rounded breeder will know some of the specialties, and know how to converse and collaborate with specialists in other parts of plant breeding.

Farming, like plant breeding, has become broader in its compass and more specialised in its technology and science. Today's crop and fiber production is much more specialised than in earlier decades. Production is done on a larger scale, and with greater sophistication of machinery and agronomic practices. And farmers are more highly educated and more sophisticated than in earlier years, even in comparison to the 1970s. Although farm sizes on the average are larger, a significant number of producers grow high value specialty products on relatively small acreages. They grow new crops such as kenaf or quinoa, or old crops in new ways (or with new adaptations of old ways), e.g., "organic " crops (crops grown without use of synthetic chemical pesticides or fertilisers) to be used for food or animal feed.

Crop agriculture can be divided into three broad categories:

(a) production of major commodity crops such as wheat, corn, and soybeans,

(b) production of crops with a smaller market (sometimes called minor crops) such as grain sorghum, sunflower, and forage grasses, and

(c) production of specialty crops such as canola (nearly out of the specialty class), kenaf, or organically grown grains, fruits and vegetables.

Plant breeders are needed for genetic improvement in all three crop categories; their goals are similar but emphases vary—yield is of predominant importance for corn, oil percentage for sunflower, and disease and insect tolerance are highly important for organically grown crops. A new sub-category is appearing, at the moment only in the major crops. Markets are beginning to develop for biochemically differentiated forms of soybeans and corn, e.g., soybeans with different oil qualities, or corn with higher oil content. Breeders talk of lowering the phytate content of feed grains to increase availability of phosphorus in animal feeds. Eventual market size of these products is unknown; some forecast that they will dominate the markets; others are think they will be less important.

PLANT BREEDING INDUSTRY

To some extent large international seed companies have different needs from regional, smaller companies. International companies need to have some breeders with language skills, or who are natives of countries where the international companies operate. But on the whole, the geographic scale of operation does not determine the kind of breeder skills that are needed; plant breeders in the USA, France, India, Zimbabwe, or Argentina all need the same basic skills and all find instant compatibility when they discuss their trade over a cup of coffee, or a beer.

Breeders of crops with "value-added" traits such as "low-sat" soybeans, or Bt corn need extra skills or more likely they need to know how to work collaboratively with scientists with special skills in biochemistry, genetic transformation, or molecular marker technology. Perhaps one might say, alternatively, that breeders who are experts in inserting and tracking valueadded genes into crop plants need to know how to work collaboratively with scientists who are skilled in field plot technique and in artful manipulation of germplasm to make farmer-acceptable varieties.

Starting with the Plant Variety Protection Act (PVPA) of 1970, and progressing to use of patents to protect germplasm of all sorts, plant breeding has become thoroughly mired in intellectual property matters. Plant breeders do not need to be lawyers but they need to know how to collaborate with lawyers who are skilled in intellectual property matters. They need to know when to protect their products, and what procedures should be followed for patenting or PVP. And even more, they need to know how to safeguard their protected products. Unauthorised use of protected materials—whether by accident or on purpose—is an inevitable byproduct of patenting and PVP.

The food production system is more integrated than it used to be. Final points of sale have more and more control over supplies, all the way down the chain to the farm. Contracts, sometimes outright ownership, and other kinds of organisation are used to give better assurance of product delivery when needed, in desired amounts, and with desired quality. Seed companies find themselves in the middle of these new chains of supply. In some cases, plant breeding departments are included as one part of a wholly-owned, vertically integrated food chain, from farm to super market, or some portion thereof.

More typically, seed companies work with contracts or carefully crafted alliances with producers and processors up and down the chain. ("Network" might be a better term than "chain", for essential connections in the food supply system are horizontal as well as vertical; a picture of the entire food supply system might resemble a fish net—or a lace curtain—more than a linked chain.) In any case, plant breeders need to know much more about the food supply system and its operators than used to be the case. Dealing with farmers alone is no longer sufficient.

TRAINING PLANT BREEDERS

Land grant universities train plant breeders, but the task is much more complicated than it might seem, from the outside.

Research vs. Teaching

The land grant universities have a dual responsibility for agriculture: research and teaching are mandated simultaneously, and teaching is subdivided into on-campus teaching of students, and off-campus teaching through the extension services. From the beginning, it has been hard for universities to strike a balance among these mandated duties that pleases everyone. Seed companies, for example, want the universities to produce graduates who can add value to the seed companies, but they also want certain kinds of research products, either finished varieties, breeding stocks, or useful new knowledge about genetics and breeding. But, not surprisingly, the seed companies don't speak with one voice; the needs of smaller regional companies may be different from those of seed companies that are part of an international agricultural chemical company. Companies dealing with corn or soybeans have a different point of view from those that produce and sell turf grass seed.

Sources of Funds

Funding for the land grant universities has changed drastically during the past two or three decades. Assured sources of funds for long term research projects (such as plant breeding) are reduced to the point that they are nearly invisible. Both federal and state funding have been reduced, countrywide, although states vary greatly with regard to how much state support is supplied, and for what purposes. Research success in the universities depends on an adequate supply of materials and facilities but it also requires an adequate supply of industrious, intellectually curious laborers, that is, university research depends heavily on the work of graduate students. Funding support for homegrown graduate students in plant breeding has dwindled or even disappeared; in many programmes the majority of graduate students are foreign nationals on stipends from their governments or from foreign aid agencies.

Nearly all biological research, including that in crop science, now depends on peer-reviewed grants, usually with a three-year life span, and no guarantee of renewal. Granting agencies place great emphasis on projects that promise to increase basic knowledge in the science, and discriminate against routine, long-term projects that merely turn out useful products, such as new plant varieties or improved germplasm for use by plant breeders.

In some states, commodity groups raise money, as through check-offs, that is ear-marked for plant breeding and allied research for that commodity. There is no nation-wide uniformity in this practice, and there are no formal guarantees that such practices will continue, even where they are in effect. Many universities now protect the products of their plant breeding and genetics programmes with plant variety protection (PVP) certificates, and patents.

Such practices were first instituted in about the 1980s. Intellectual property rights and their royalties can bring income to plant breeding programmes, but such income can be less than it seems, and irregular in nature. When all costs are taken into account, expenses of administering intellectual property rights programmes at the universities nearly always exceed the income from patents and PVP. An occasional patent or PVP certificate can earn large sums of money for a period of time, but to depend on royalties to support a university plant breeding programme is somewhat like trying to make a living by playing the state lottery.

Responsibility to General Public vs. Special Clients

The land grant universities, in addition to their two mandated duties of teaching and research, have two categories of mandated clients: the general public and special groups that can use their products directly. For the general public, the universities need to do research and teaching that (for example) will help agriculture to be more sustainable and environmentally responsible. For special clients, such as farmers or plant breeding companies, the universities are expected to do research and teaching that helps the clients directly, e.g., they should give farmers information that will increase profitability of soybean production, or provide seed companies with research results, or germplasm—

r well-trained breeders—that will increase the seed companies' ability to serve the farmer, at a profit. Inevitably the universities are caught between opposing points of view: those on the one hand, who say that the universities are doing research that is irrelevant to the needs of agricultural producers or agribusiness companies, and those on the other hand, who say that the universities are nothing but pawns in the hands of large farmers or agribusiness, spending public tax moneys for private gain of the privileged few. And seed companies themselves, as already noted, do not necessarily agree on which kind of training, or what research problems and products, are proper for the land grant universities.

Role of Industry

Industry necessarily must do some training of its plant breeders, at the least in company research procedures and sometimes in use of proprietary skills and knowledge. Often, breeders come equipped with knowledge about breeding one crop but are assigned to a different crop; the company needs to train the breeder in the biology, the genetics, and the special selection needs for that crop. But all skills and all knowledge can become dated, usually because of advancing research and scholarship in the universities. Especially in recent years, with the explosion in scientific knowledge and technology, seed companies find that their breeders are out of touch with new knowledge and new techniques. There are two ways to update the breeding staff. One way is to fire the older, less productive breeders and hire new graduates with up-to-date skills and knowledge. Another way is to send the breeders back to school via a sabbatical, or through collaborative research and study with university staff. Both ways have been tried, but it would appear that the first method is more common, at this time. In either case, the universities are essential purveyors of education and training.

Role of Non-land Grant Universities

Especially since biotechnology became a part of plant breeding, graduates of non-land grant universities have been hired by seed companies. In most cases these graduates have had minimal knowledge of agriculture and classical plant breeding but they were enthusiastic about learning how to use their skills to improve plant varieties. When given the right atmosphere to learn and interact with field breeders, the non-land grant graduates have been valuable additions to the plant breeding team. Their numbers are likely to increase, proportionately, in the future.

Biotechnology in Medical Research

Medical research today has direct application to plant breeding; genomics as developed for mammals including humans can identify genes with clear homologies to plant genes. Functions of the animal genes may or may not be markedly different from their plant homologues, but the wealth of information coming from the well-financed medical research establishment increasingly will be useful to plant breeders, even though most of the new knowledge will not have practical application for field breeding projects in the near term.

Educational Needs

Plant breeders have always needed education and skills that go well beyond breeding and genetics. They need to know good business practices, practical economics, and they must be able to express themselves clearly and accurately in speaking and writing. In order to interact successfully with their now broader audience (including, nowadays, energetic critics of many products of modern plant breeding), they need to understand sociology and societal mores and norms. Skill in one or two foreign languages will always be useful, especially as foreign trade becomes more and more pervasive. These skills are needed by breeders in small as well as large seed companies, in regional as well as international companies.

The first professional plant breeders only rarely had received the Ph.D. degree, and many of them had no college education at all. They were successful in spite of (or maybe because of?) their lack of higher education. As years went by, new plant breeders for industry usually were required to have at least an M.S. degree, and then a Ph. D. in plant breeding and genetics. Now, seed companies often want people who have post-doctoral experience, and they prefer that it be in a field different from the one in which they took their doctorate. But at the same time, needs are increasing for top quality people with degrees at about the M.S. level, to do many of the skilled operations with the new kinds of highly technical machinery and scientific apparatus. A high school education is insufficient for this work, and a general purpose M.S. in agriculture or biology may be insufficient, as well.

TRADITIONAL PLANT BREEDING AND BIOTECHNOLOGY

Some people believe we are at a crossroads in regard to the kinds of training needed for plant breeding of the future. Breeders trained in classical field breeding are contrasted to breeders with primary training in biotechnology. What will be the needs for either kind of breeder in the future, or will plant breeding companies require breeders that can do both kinds of work? It will help to look at various aspects of breeding, making guesses as to what kind of breeding will be most useful in the future.

Yield Potential

Yield potential, defined as maximum yield that can be achieved in a given environment when constraints to yield have been minimised, still is best increased by use of traditional breeding methods, that is, by crossing best with best, and then selecting rigorously over many environments and years, in the locale where the varieties will be grown. A combination of art and experience, using trial and error, is still the best way to develop varieties with maximum yield potential.

Yield Stability

Yield stability is required for expression of maximum yield potential, assuming that one wants varieties that outyield all others, year in and year out. Therefore, developing varieties with maximum yield stability requires the same skills—art and experience—that are useful for developing varieties with maximum yield potential.

Disease and Insect Resistance

Genetic transformation, with Bt as its prize example, seems to have sidelined traditional methods of breeding for disease and insect resistance. Multifactorial kinds of resistance, or use of single genes from the same or closely related species, seems to be old-fashioned. But use of molecular markers will enable faster backcrossing of single genes for resistance and allow breeders to manipulate them more easily than before, therefore it seems likely that there will be much use of same-species sources of resistance for many years to come.

The probability of pathogen or insect development of biotypes able to overcome single gene resistance seems to have little bearing on the source of the gene; transformed resistance genes probably will become useless just about as fast as same-species genes for resistance. Thus, after initial enthusiasm for the silver bullets of foreign genes for major resistance, breeding will settle down into evolution of optimum mixes of multifactorial and single gene resistance, just as it did before the advent of genetic transformation. But molecular biology gradually will unlock many secrets of ways to develop longer lasting and superior pest resistance. Major gains in developing such kinds of resistance will depend on plant breeders, entomologists, and plant pathologists with great skill—or good collaborators—in biotechnology; these specialists will be in great demand.

Special Traits, Grain Quality, Composition

Biotechnology also will give a proportionately large contribution to advances in breeding for special chemical and quality traits in grains, forages, and fiber crops. But such changes will need to be incorporated into elite genetic backgrounds and thorough testing will be required to be sure that undesired agronomic traits have not been introduced inadvertently by the transformation process, through linkage drag, or even due to pleiotropic effects of the new genes. Traditional and molecular breeders will need to work hand in hand to provide reliable products, prevent undesirable surprises, and do so with maximum speed, in order to keep ahead of the competition. And of course in the background and in the foreground will be the need to be sure that genetically altered varieties are safe for human and animal consumption, thus ties with the medical profession and nutritionists will be essential as well. People with such training will be needed, as part of the plant breeding team of the future.

Needs of Seed Ccompanies

As one might expect, the size of the seed company, whether or not it does its own breeding in-house, and the nature of its major seed products, governed the needs that the seed company administrators and researchers felt to be most pressing. Breeders of minor crops saw a general reduction in university emphasis on their crops. Trained breeders, as well as finished crop varieties or inbred lines suited to their businesses, are becoming scarce and scarcer.

As a consequence, some of the companies are striking up alliances with entrepreneurial breeding firms, such that one company specialises in development of varieties that are leased out to other companies that specialise in production and sales. In this way, they are following the route long since taken by small hybrid seed corn companies in the Corn Belt. Foundation seed companies develop inbreds and recommend hybrid combinations; other companies lease the inbreds and produce and sell their own proprietary hybrids. But in some cases, the market is too small for even specialised small breeding companies. In such cases, the seed companies say they must have products from the public sector, or they are out of business. The future does not look good for future advance in such "orphan" crops.

Larger seed companies are more likely to want breeders with experience in biotechnology, or biotechnologists with knowledge of plant breeding. They now are looking for people with not only a Ph.D. but also with post-doctoral experience. Their representatives often said that the supply of well-trained biotechnologists is adequate, but they have to look hard to find good people trained in field breeding. So in this respect, people from small and large companies agree: the supply of top-notch people with training in field breeding is tight or even too small for comfort. Representatives from all companies agreed that today's and certainly tomorrow's breeders cannot possibly have mastered all the skills that must be applied in a modern plant breeding programme, but they at least must be acquainted with, and have some knowledge about, the full range of sciences and technologies. More importantly they must be good listeners; they must be able to hear and understand what the other specialists have to say about their contribution to the total plant breeding programme.

When one tries to describe the skills and training that make a wellprepared plant breeder, it is clear that the answer for today will not be the answer for tomorrow. As one company researcher said, "Plant breeding needs a new definition; it is not like it was. Plant breeders continually will need to learn new skills, either in the company or in the universities."

University administrators agreed in acknowledging that when biotechnology first began to promise great things for plant breeding (about 15 years ago), too much emphasis was placed on support of laboratory research and training, and field-oriented plant breeding was under-supported. Now, they say, they are placing more emphasis on training breeders in field-oriented, classical kinds of plant breeding, and feel that the balance is about right. The administrators are worried about funding of plant breeding research and training, however.

Funding support seems to evaporate in every direction they turn. For example, people in sciences that give essential support to plant breeding research (e.g., statistics, entomology, plant pathology) no longer are willing to give collaborative support to plant breeding programmes. They are opting out for the usual reason—granting agencies won't support such activity because it is too routine and practical; it is not basic research. Breeders are forced to develop their own support in such fields as entomology, statistics, and plant pathology, and this depletes their already hard-pressed research funding base.

Equally important, and directly to the point of this discussion, is that funds for graduate assistantships are seriously inadequate, particularly for training in classical plant breeding. The administrators said that in a few instances private industry is ear-marking grants for such training, but that much more help is needed. Some administrators said that it is time to consider training breeders in collaboration with industry, that industry could furnish materials and facilities for the students in ways that no longer can be done at the universities.

Industry would need to work out ways to protect confidential information and germplasm, but since they already are working out agreements on the same topics for collaborative work with other members of industry, it might be fairly easy to do the same for work with students and faculty of the universities. The university administrators are worried about how to fund research and training in breeding of the minor crops. Some suggested that the universities need to specialise and cooperate in research and training for the minor crops, dividing up the minor crops so that one university (for example) does oat breeding for several states, another handles alfalfa breeding, and so on. In some cases this kind of collaboration has begun already, but much more is needed.

It is very hard for the land grant universities to break out of the habit of 16 serving only their own state, especially when they have to depend on state legislators for a large share of their total income. None of the administrators seemed to fear that industry was unduly influencing their research and training programmes. If anything, they said that industry was not talking to them enough about its needs; the administrators had to find out for themselves what needed doing, to properly serve the needs of agricultural industries.

The university professors were much more pessimistic than the administrators, about present and future prospects for training adequate numbers of plant breeders in classical techniques, but they agreed with the administrators, in saying that funding is no longer available for classical plant breeding, or for any kind of project that requires many years of single-minded effort to achieve its goals. The professors were not happy with the current tendency, as they see it, for new faculty to be hired to do both field breeding and biotechnology in support of that breeding. They believe that it is a mistake to expect faculty to be both field breeders and laboratory investigators; they say that either of the specialties is a full-time job and that anyone who tries to do both will end up doing neither.

An extremely disturbing complaint from some of the professors was that the quality as well as the number of applicants for graduate study in plant breeding is low, and going lower. They said that the best of the science-oriented undergraduates in agriculture are going into laboratory sciences, not fieldoriented plant breeding. And many top students are leaving college with an undergraduate degree, in order to take a high-paying sales job with agribusiness. In either case, they are going where the money is. The plant breeder professors say that as a consequence, no one today can turn out a succession of field breeders in numbers and quality that were produced by now-retired plant breeders in the previous two or three decades. They say that industry is looking for breeders but they cannot supply the demand.

At the same time, they wonder how serious the seed industry really is, about wanting field-oriented plant breeders, as they look at recent layoffs of mid-career field breeders with no corresponding lay-offs of biotechnology staff. At best, the professors seem to get little encouragement or help from industry, to enable them to supply industry's needs for plant breeders of the future.

STATUS OF RESEARCH AND APPLICATION OF CCROP BIOTECHNOLOGIES

FAO and its Governing Bodies recognize the role biotechnology can play in augmenting agricultural production when properly integrated with other technologies. Member countries look to FAO's assistance in strengthening their institutions through provision of technical, legal and policy advice as well as promotion of information exchange. Information on biotechnology activities in developing countries is scarce and this has prompted FAO to develop an inventory of plant biotechnology products and techniques in use or in the pipeline in developing countries.

The inventory has been compiled and organized into a searchable online database called the FAO Biotechnology In Developing Countries Database (FAO-BioDeC). The FAO Committee on Agriculture (COAG), whose main task is to review and provide advice on FAO's programme of work in food and agriculture, including on selected international development issues, stressed in 1999 that biotechnology, when coupled with other technologies, offers considerable potential and opportunity for new solutions to some of the old problems hindering sustainable rural development and achievement of food security, but also notes that it is an area where there is a growing gap between developing and developed countries. To take advantage of the biotechnology promise, many FAO member countries need assistance in strengthening their overall capabilities in research and development, and at the policy and regulation level.

The member countries have in particular recommended that FAO provide, on request, policy advice for biotechnology issues related to food and agriculture; promote information exchange, and provide technical and legal assistance and advice to its members. In addition, member countries recommended that FAO develop a strategic approach with emphasis on the coordination of a crosssectorial programme on biotechnology. Priority Areas for Interdisciplinary Action (PAIAs) have been identified within FAO's strategic framework for 2000-2015. An Interdepartmental Working Group on Biotechnology (IDWGB) was established in 1999 to guide and coordinate FAO's strategic approach under the new PAIA on biotechnology.

FAO-BIODEC

For the purpose of this inventory, biotechnology has been defined as "any technological application that uses biological systems, living organisms, or derivatives thereof, to make or modify products or processes for specific use". This definition can be interpreted in a wide sense to cover not only recombinant DNA technologies such as genetic engineering which result in transgenic organisms, but also to include other forms of nontransgenic biotechnology R&D such as genomics, marker-assisted breeding, micropropagation and agricultural diagnostics. At this stage, the FAO-BioDeC includes about 2000 entries from 71 developing countries, including countries with economies in transition. The database in this initial phase is limited to research, testing and commercialization of specific crop technologies and products in developing countries. No quantitative information is available with regard to the human capacity or funding involved. It does not cover activities carried out in developed countries even if they are meant for subsequent use or adoption in developing countries, nor does it cover research being carried out at international research centres, such as those that are part of the Consultative Group on International Agricultural Research (CGIAR), located in developing countries.

To compile the FAO-BioDeC a number of data sources have been consulted. In particular, information on plant biotechnology products and techniques gathered from a survey undertaken in Latin America by the International Service for National Agricultural Research (ISNAR) and from country biotechnology status assessment reports undertaken by FAO in South East Asia, Africa, and transition countries in Eastern Europe. The initial biotechnology application data obtained has been classified on a country/regional/continental basis, by species, trait analysed or technique used, and by whether the application is in the research phase, field testing phase or already commercially released (or routinely utilised in the case of non-GM technologies).

Microbial Products For Agriculture

Microbiological techniques are central to the production, harvesting and use of many agricultural products. The use of microbial techniques includes the design or delivery of microorganisms for the control of pests (biocontrol agents or biopesticides), as fertilisers (biofertilisers) and for fermentation and food processing techniques. From data in the FAO-BioDeC, it seems that research towards the scientific development of biocontrol agents (biopesticides), biofertilisers and fermentation and food processing microbial techniques, is at the early stages in Africa and Asia, (with the exception of China and India), while Latin American countries are already using advanced techniques on a more routine basis and some of the results obtained are already being tested. It should be borne in mind that many production and post-production processes already in use by farmers have a microbiological basis and that the extent of use of microbiological techniques in agriculture is grossly under reported by focusing only on the use of microbiological techniques which emanated from 'modern' laboratories, rather than also on the use of microbiological techniques developed as indigenous or local knowledge of farmers and consumers.

Microbial Agents for the Biocontrol of Pests

Although considerable use of classical pesticides persists, in some countries there is a trend towards using newer pesticides that are more selective, less toxic to humans and the environment, and require less application per hectare to be effective. A small but growing percentage of these are biopesticides, including microbial pesticides like *Bacillus thuringiensis* (Bt), and biocontrol agents such as pheromones, growth regulators and hormones. In addition, there is growing acceptance of use of such alternative pest control agents in various forms of Integrated Pest Management (IPM). Biocontrol agents, or biopesticides, range from the classic Bacillus thuringiensis (Bt) to Trichoderma, Verticillium, Bauveria, *Bacillus subtilis* to plant extracts, entomophagic nematodes or entomopathogenic viruses, such as nucleopolyhedrosis virus (NPV).

Biofertilisers

Biological nitrogen fixation (BNF) refers to the process of microorganisms fixing atmospheric nitrogen, mostly within subsoil plant nodules, and making it available for assimilation by plants. Nitrogen availability is a key limiting factor in crop production. *Rhizobia* are the most studied and important genera of nitrogen fixing bacteria, but also a number of endophytic bacteria are now known as nitrogen fixing organisms. Use of biofertilisers, such as *Rhizobium* or other symbiotic and non-symbiotic species for atmospheric nitrogen fixation represents a more environmentally-friendly alternative to chemically generated fertilisers. Other microorganisms, such as *Mycorrhiza*, are active in establishing symbiosis with cultivated plants and forest trees, and facilitate phosphorus uptake. Inoculation with these fungi has proven to be an efficient way to substitute or complement phosphorus-based chemical fertilisation.

The only reported research on biofertilisers in the Near East is research into *Rhizobium* in food legumes in Morocco. In Latin America, techniques for increasing nitrogen fixation in soybean, bean and cowpea are being tested in Venezuela, and researched in rice in Brazil, where arbuscular mycorrhiza are also being tested in native trees. Argentina has started nitrogen-fixation research on unspecified microorganisms. In Asia, India is researching *Rhizobium* with high capacity for nitrogen fixation, and China is studying *Rhizobium* in rice and maize. Indonesia is investigating vascular arbuscular mycorrhiza, the Philippines mycorrhiza and *Rhizobium* in forest species, and Viet Nam is experimenting with *Rhizobium* strains for the Mekong delta soils.

In Africa, research into biofertilisers for sorghum is ongoing in Ethiopia, Kenya and Zimbabwe, for cowpea in Cameroon, for groundnut and bambara groundnut in Madagascar, and for rice in Rwanda, with unspecified work in Burkina Faso, Cote d'Ivoire, the Democratic Republic of the Congo, Kenya, Rwanda and Senegal. The UNESCO Microbiological Resources Centre (MIRCEN) project at the University of Nairobi in Kenya has, since 1981, developed a *Rhizobium* inoculant known as *BIOFIX*, currently the main inoculant available on the local market. In Eastern Europe, Armenia and the Republic of Moldova are testing *Azotobacter* and *Rhizobium*.

Fermentation Technology and Food Processing

Biotechnology in the food processing sector targets the selection and improvement of microorganisms with the objectives of improving process control, yields and efficiency as well as the quality, safety and consistency of bioprocessed products. These microorganisms are used in fermentation; the process of bioconversion of organic substances by microorganisms and/or enzymes (complex proteins) of microbial, plant or animal origin. Fermentation is one of the oldest forms of food processing which is applied globally. In developing countries, fermented foods are produced primarily at the household and village level, where they find wide consumer acceptance. Food fermentations contribute substantially to food safety and food security, particularly in the rural areas of many developing countries. The extent of application of biotechnologies such as fermentation technology and various food processing techniques is immense and difficult to monitor using a database such as the FAO-BioDeC. Nevertheless, it indicates that fermentation technology and food processing is quite commercialised in Latin America for sorghum in Brazil, where food enzymes are also being tested in several species of grains and vegetables, and researched in sugar cane.

In Eastern Europe, *Lactobacillus, Acetobacter* and *Saccharomyces* are reported as being used commercially in Armenia, Georgia and the Republic of Moldova , where *Bacillus subtilis,* is also being used. *Actinomycetes* are being tested in Azerbaijan, and *Bacillus vulgaris* researched in Armenia. In Africa, research phase initiatives are underway in the fermentation of banana in Kenya and Zimbabwe, cowpea in Cameroon, cassava in Nigeria, with unspecified research in Burkina Faso. In Asia, Indonesia is researching lactic acid bacteria, dextranase, xanthan gum production and soybean fermentation. China and Indonesia are reported to be working on the use of the phytase enzyme in improving animal feed.

Environmental Biotechnology

Environmental biotechnology can be defined as the development, use and regulation of biological systems for remediation of contaminated environments (land, air, water), and for environment-friendly processes (green manufacturing technologies and sustainable development). The FAO-BioDeC reveals few environmental biotechnology initiatives of relevance to food and agriculture underway in the countries surveyed. However, it is reported that microbiologial techniques for environmental management are being investigated in Indonesia, for biological waste and water treatment and bioconversion of solasodine, and in Thailand for biodegradable plastics and waste treatments. In Eastern Europe, Azerbaijan is testing *Basidiomycetes, Deyteromycetes,* and *Fungi imperfecti* and Georgia is researching *Mycobacteria, Halobacterium halovium,* and *Aspergillus terreus.* In Latin America, Chile is testing *Thermomyces lanuginosus* for cellulase production, and bacterial biofilm for bioremediation.

Applied Plant Cell Biology Techniques

Micropropagation

Micropropagation is the use of tissue culture methods to propagate plants. Using micropropagation, millions of new (clonal) plants can potentially be derived from a single plant. Micropropagation encompasses a range of tissue culture techniques for propagation of plant species. In essence, tissue from a plant (explant) is isolated to create a sterile culture of that species *in vitro.* Once a culture is stabilised and growing well *in vitro,* multiplication of the tissue or regeneration of entire plants can be carried out. Shoots (tips, nodes or internodes) and leaf pieces are commonly used but cultures can be generated from many different tissues. Juvenile tissues generally respond best. This method of cultivation of plant material is used for:

1. rapid, large-scale, year round production of desired horticultural varieties (e.g. many of the popular miniature rose varieties are produced this way);
2. propagation of plant species that are difficult to grow from seed;

3. production of genetically uniform plant material ("clones");
4. development of plant culture systems that can be used for genetic transformation, e.g. to introduce disease resistance; and
5. production of disease-free plant material (e.g. potato microtubers).

The benefits of plant tissue culture propagation include potentially unlimited multiplication of superior plant lines or elite individuals, avoidance of contamination with pathogens, production of true-to-type multiplication material of desirable plant lines suited for indefinite storage through cryopreservation, or for long-term maintenance of propagule inventories. In comparison to other plant propagation techniques, the major limitation of the application of this technology is the need for technically skilled labourers and some essential equipment. It is therefore commercially applied to high value added crops, which are worth the necessary investment.

Anther and Pollen Culture

Anther culture involves the aseptic culture of immature anthers to generate fertile haploid plants from microspores. The production of haploid plants through anther culture is widely used for breeding purposes, as an alternative to the numerous cycles of inbreeding or backcrossing usually needed to obtain pure lines in conventional breeding. The success achieved with anther culture has led to the development of microspore culture. This involves the isolation of the microspores from the anthers, culturing them in specialised media and subsequent regeneration of fertile homozygous plants. Furthermore, isolated microspores are very attractive for protoplast isolation and applications aiming at transformation as they are unicellular and transgenic homozygous plants could be provided in a comparatively short time. *In vitro* anther culture is now routinely used for improving some vegetable crops such as asparagus, sweet pepper, eggplant, watermelon and *Brassica* vegetables.

In addition, anther culture is being increasingly used in cereal crop improvement both as a source of haploids and new genetic variation. Isolated microspore culture has been successfully carried out with some *Brassica* vegetables such as cabbage, broccoli and Chinese cabbage-petsai and pakchoi. Most genotypes respond better to isolated microspore culture and embryo yield is generally higher than with anther culture. Therefore, isolated microspore culture has been preferred as a breeding tool and as an experimental system for various genetic manipulations. A Chinese cabbage breeding programme using this technique is underway in the Beijing Vegetable Research Center.

In Asia, three countries are reportedly using anther culture. In China, for example anther culture is being applied in different species such as wheat, rice, maize, rubber, hot pepper, poplar and sugar beet. Bangladesh is conducting trials on anther culture derived rice with salt tolerance and anther culture research on mulberry, maize, jute barley and tea, and Nepal has research on rice and wheat. In the Near East, anther culture research emphasis is on cereals. An anther culture generated durum wheat variety has reached the commercial phase in Tunisia, and anther culture techniques are being researched in Morocco, where an anther cultured bread wheat is in commercial use. Anther culture techniques are being tested in the Sudan, and applied to barley in the Islamic Republic of Iran, Iraq, Morocco and the Syrian Arab Republic. Wheat is also being researched regarding anther culture in Eastern Europe in the former Yugoslav Republic of Macedonia and Serbia and Montenegro.

In Latin America, varieties obtained through anther culture of unspecified fruits, cereals and ornamentals have reached the commercial stage in Brazil. Advanced testing of anther culture derived varieties of 'many plants' is reported in Costa Rica, and in Venezuela cocoa, rice, sugar cane, papaya, mango and bamboo are also being tested. Anther culture derived rice varieties are under advanced testing also in Argentina, and in Brazil, where this technique is specifically used in breeding for blast resistance. In Chile, wheat, barley and rice, are in the research phase. Other anther culture research is carried out on maize and onion in Argentina, wheat in Ecuador and asparagus in Peru. In Africa, anther culture research has started on banana in Cameroon and on sorghum and rice in Mali.

Embryo Rescue

In embryo rescue/culture, the embryo is removed before seed abortion occurs and is grown outside the parent plant to produce a new plant to enable crosses to be made between species which would not normally be sexually compatible. *In vitro* embryo rescue techniques are therefore often used to rescue plant embryos from aborting progeny seeds that result when two distantly related plants (e.g. two species) are crossed together. Such 'wide crosses' are often desirable to transfer genetic traits from the secondary and tertiary genepools (i.e. crop wild relatives) to the cultivated primary genepools of crop plants. This technique is used in breeding of many crop species and allowed synthesis of triticale, a new hybrid species resulting from the cross between rye and wheat.

In the Near East, embryo rescue techniques are being developed in Algeria, Morocco and the Sudan. In Eastern Europe, embryo culture research is underway on sweet and sour cherry in the former Yugoslav Republic of Macedonia, sunflower in Yugoslavia, and maize in the Republic of Moldova. Only in Latin America has extensive use been made of the embryo culture technique, where it has given rise to commercially viable new varieties in banana, citrus, papaya, fruits, cereals and ornamentals in Brazil. New embryo rescue-derived varieties are tested in maize, *Celtis tala, Erthytrina crista galli, Acacia caven* and *Elymus spp*. in Argentina, in grape in Chile, in orchids in Costa Rica and in rice, sugar cane, papaya, *Solanum nigrum*, Yantén, *Spathiphyllum* sp., *Aster* sp., and gloxinia in

Venezuela. Research is at an early stage in Africa, on yam in Nigeria and unspecified work is being carried out in Cameroon, and also in Asia, where the only embryo culture report is of work on rice and orchid in Bangladesh.

In vitro Regeneration and Somaclonal Variation

Plant regeneration from cell cultures is central to the application of gene transfer techniques such as biolistics and *Agrobacterium*-mediated transformation. Not all plants are readily amenable to *in vitro* regeneration and there is thus a need to continuously develop regeneration protocols for the recalcitrant species if they are to benefit from genetic modification technologies. In vitro regeneration usually results in high genetic and phenotypic variability in individuals derived from cultures, which is called somatic variation. Somatic variation can be beneficial in crop improvement especially on traits for which somaclonal mutants can be enriched during *in vitro* culture, including resistance to disease pathotoxins, herbicides and tolerance to environmental or chemical stress, as well as for increased production of secondary metabolites.

Advanced *in vitro* regeneration work (protocol testing) is reported in *Acacia caven, Erythrina crista galli*, potato, sweet potato, cassava and garlic in Argentina, in blueberry in Chile, in cassava, aroids, *Musa* sp., *Coffea* sp., banana and plantains in Costa Rica, in *Ullucus tuberosus, Oxalis tuberosa, Tropaeolum tuberosum, Arracacia xanthorrhiza, Mirabilis expansa, Canna edulis*, and manihot in Ecuador and in tropical fruits, roots and tubers, cassava, potato, *Musa, Solanum nigrum*, Yantén, *Spathiphyllum* sp., *Aster* sp., and gloxinia in Venezuela. Research is in the laboratory stage (protocol development) on *Parkinsonia aculeate, Elymus* sp, and *Ilex paraguariensis* in Argentina, red clover and garlic in Chile, and on *Mycosphaerella fijiensis, Rosellinia spp.* and selected plant species in Costa Rica. In Asia, the only reported initiative is research on somaclonal variation in mulberry in Bangladesh.

In vitro Germplasm Conservation and Cryopreservation

Biotechnology can play an important role in the conservation of germplasm of many crop species, whether the germplasm is maintained as vegetative or seed propagules. *In vitro* conservation protects germplasm from possible contamination with pathogenic agents and preserves the genetic identity of the stored material. Germplasm regeneration techniques (*in vitro* or *in vivo*) coupled to cryopreservation protocols ensure the long-term, safe storage of much of the world's germplasm, even if this technique is well established only for a number of plant species.

The FAO-BioDeC include information on the use of *in vitro* conservation and cryopreservation. In this regard, the only work reported in Asia is experiments in Viet Nam on sugar cane and in Bangladesh on orchid, bamboo and hybrid *Acacia*. Similarly in Africa, reported work is only in the experimental phase, concentrating on root crops, cassava in Cameroon, Ghana and Nigeria, yams in Cameroon, Malawi and Nigeria, sweet potato and potato in Kenya, also on banana in Ghana, Malawi, Nigeria and Uganda, with unspecified activities in Cote d'Ivoire. In Latin America, use in cassava and violet is reported in Brazil, where experiments are also underway on sugar cane. Chile has developed cryopreservation of *Fragaria chiloensis* at the laboratory stage.

In the Near East, only Egypt has any activity, in the laboratory, on potato conservation. In Eastern Europe, the former Yugoslav Republic of Macedonia is conducting trials on techniques for *in vitro* conservation of grape, apple, strawberry and cherry, and Yugoslavia for sunflower and potato. Experimental work is ongoing for potato in Albania and Bosnia and Herzegovina, also on *Populus, Genthiana sp.,* and 30 wild species in Bosnia and Herzegovina and on *Salvia hydrangea, Bryonia alba, Atropa belladonna, Rubia tinctorum, Hypericum perforatum,* and *Melissa officinalis* in Armenia and grapevine in Azerbaijan.

Molecular Marker Techniques

Molecular marker techniques (employing RFLP, RAPD, microsatellites, AFLP, SNP and other kinds of marker systems) represent a rapidly evolving suite of powerful research tools for the characterisation and management of genetic polymorphism (variation and diversity) in plant breeding and germplasm characterisation programmes. Where they are available and cost-effective, molecular markers can have a wide number of applications in plant breeding, the most commonly considered being marker-assisted selection (MAS). The FAOBioDeC suggests that the earliest generation forms of DNA-based molecular markers (RAPD, RFLP) are more widely used than the more recently developed types of molecular markers (microsatellite and AFLP markers).

MAS is based on the identification and use of markers which are linked to the gene(s) controlling the trait of interest. By virtue of that linkage, selection may be applied to the marker itself. The advantage consists in the opportunity of speeding up the application of the selection procedure. For instance, a character which is expressed only at the mature-plant stage, may be selected at the plantlet stage if selection is applied to a molecular marker. Also, selection may be applied simultaneously to more than one character, and selection for a resistance gene can be carried out without needing to expose the plant to the pest, pathogen or deleterious agent. Finally, when there is linkage between a molecular marker and a quantitative trait locus (QTL), selection may become more efficient and rapid. The construction of detailed genetic and molecular maps of the genome of the species of interest is a prerequisite for most forms of MAS. However, the current cost of the application of these techniques is significant, and the choice of one technique rather than others may be dictated by cost factors. There are still very few examples of crop varieties in farmers' fields which have been developed based on MAS, largely because of the currently prohibitive cost of incorporating large-scale MAS into the budgets of most plant breeding programmes.

In addition to MAS, molecular markers can be used in germplasm characterisation. Compared to morphological and protein markers, DNA-based genetic markers are often considered to be the most useful for genetic diversity studies because they are highly polymorphic and heritable (their expression is not affected by environmental variability). The FAO-BioDeC indicates that molecular markers are being extensively used in Latin America with 93 trials and 165 molecular marker projects at the research phase in nine countries.

The assessment of the genetic diversity of Andean local roots and tubers in Latin America using molecular markers is an interesting development. Species reported to be included in molecular marker programmes in these countries are sugar cane, rice, cocoa, banana, bean and maize. The survey also indicates that most countries in Asia are undertaking a wide spectrum of crop research using molecular markers. Molecular marker-related research activities in Africa are reported to be underway in only a few of the countries, such as Ethiopia, Nigeria, South Africa and Zimbabwe; the range of Africa crops under study with molecular markers, however, is very wide: from traditional commodities to tropical fruits.

Dna and Immuno-Diagnostic Techniques

Many crop diseases are difficult to diagnose, especially at the earliest stages of infection by the pathogen. In particular, many diseases caused by viruses can exhibit similar symptoms and therefore it is difficult to identify the causative virus. Knowledge of the nature of the pathogen can be used to develop and apply proper management practices. For instance, some viruses are seed transmitted whereas others are not, or some bacterial pathogens can be managed by changing the growing environment. The development of cheap diagnostic techniques could assist decision-making in relation to pest and disease management.

The development of diagnostic kits, such as enzyme linked immunosorbent assay (ELISA), and other molecular assays, can enable the precise identification of viruses, bacteria and other disease-causing agents, and is now an established tool in disease management in many farming systems. Diagnostic assays have also been developed to identify a wide range of other organisms, chemicals (such as undesirable by-products, e.g. aflatoxin), or impurities involved with food quality.

The FAO-BioDeC suggests that ELISA is only in commercial use in Eastern Europe, although this is unlikely to be unique to Eastern Europe as ELISA is undoubtedly used in plantation farming in many other regions. Most of the use of ELISA reported in Eastern Europe is to detect plant viruses (e.g. PLRV, PVX, PVA, PVM, PVY) in crops such as potato, fruits and strawberries. In the Near East, all ELISA-related work is reported to be in the experimental phase, with Iraq, Morocco and the Syrian Arab Republic working on potato, Morocco also on sugar cane, Tunisia on grapevine and Egypt on detection of

ZYMV in cucurbits. In Africa, use of ELISA for cassava diagnostics is under study in Malawi, Nigeria, Uganda and Zimbabwe, for cowpea in Cameroon and Nigeria, and maize in South Africa and Zimbabwe. In Latin America, Brazil is testing ELISA in sugar cane, potato, plum, unspecified *Solanaceae* and soybean, and researching cucurbits, *Solanaceae*, and *Rhizobium* in association with beans and forage and tree legumes, Chile is researching potato, tomato, tobacco and grapevine, Paraguay is testing ELISA in citrus, and Peru is testing potato.

DNA diagnostics are also a powerful technique for identification of pathogens and other organisms in agriculture. Most DNA diagnostics are now based on the use of the polymerase chain reaction (PCR), a common research tool used in most molecular biology laboratories worldwide, which can be used to specifically amplify segments of DNA. Most PCR techniques require the use of the enzyme *Taq* polymerase which until recently was protected by patents requiring that any commercial use of the PCR technique would have to pay royalties to the holders of the *Taq* polymerase patent. The enzyme *Taq* polymerase is now in the process of becoming a 'generic' biochemical reagent which will substantially reduce the cost of PCR applications in research and commerce. At present, most of the reported work on PCR diagnostics is in the experimental phase. In the Near East, Egypt is working on PCR analysis of ZYMV in cucurbits and TYLCV in tomato.

In Europe, Yugoslavia is researching PCR for PPV in plum, ASPV in apple and PVYV in pear. Armenia is working on diagnostic techniques for disease of tobacco, potato, linen, wheat and other crops, Azerbaijan on chickpea, tobacco and other crops, and the Republic of Moldova on maize, tobacco and other crops. In Africa, maize disease diagnostics is the subject of PCR studies in South Africa, Uganda and Zimbabwe, while diagnosis tools are studied for diseases of cowpea in Cameroon and Nigeria, cassava in South Africa and unspecified work is underway in Burkina Faso and Ghana.

In Latin America, Brazil is researching on PCR for unspecified uses in sugar cane, bean, rice, tomato, carrot, unspecified Solanaceae and banana and is conducting experiments on sweet potato, garlic, apple, genome sequencing of *Herbaspirillum* in soybean, *Rhizobium* in association with beans and forage and tree legumes, peanut and wild related species, studies on fungi, eucalyptus, Brazilian forest trees, soybean, coffee, sugar cane, *Xylella fastidiosa*, and heart of palm for food quality. Chile is researching molecular diagnostics for pathogens of potato, tomato, tobacco and grapevine, and Peru of potato.

Transgenic Crop Varieties Resistant To Pathogens

Most pathogenresistance development programmes oriented towards the development of genetically modified (GM) plant varieties, are concentrated on generating resistance to viruses and fungi, particularly in Asia and Latin America, with very little activity devoted to developing transgenic crop varieties resistant to bacteria. The current emphasis on viruses most likely reflects the relative simplicity of viral genomes, and the generic 'proof of concept' of virus-derived strategies such as coat protein and replicase-mediated resistance in generating virus-resistant transgenic varieties. Transgene-mediated resistance strategies against fungi are still in their infancy.

Overall, it is unclear from the available data in the inventory why transgenic approaches are more predominant for fungal compared to bacterial pathogens, or why some regions report more research initiatives than others. Although planting disease-resistant varieties is one of the better ways of combating viruses, bacterial and fungal plant pathogens, as with all biotic stresses with the capacity to mutate and evolve, it would be prudent to develop resistance management strategies to limit the incidence of selection for resistance-breaking pathogen isolates.

Development of Transgenic Crop Varieties Resistant to Viral Diseases

In the case of virus resistance, conventional strategies to control viral diseases are limited to the production of virus-free propagation material and to the control of insects transmitting virus pests. While some crop gene pools harbour resistance to viruses, there are crop gene pools which are completely lacking in resistance against key virus pathogens. In Africa, only three crops, sweet potato, potato and maize, have so far been targeted for transgene-mediated virus resistance.

The FAO-BioDeC indicates only two transgenic varieties that to date have been tested in a field trial, namely, a sweet potato variety for resistance to sweet potato feathery mottle virus (SPFMV) in Kenya and a potato variety for resistance to potato leaf roll virus (PLRV) in South Africa. The FAO-BioDeC currently reports only three other research initiatives in an experimental phase in South Africa, namely, the development of potato for resistance to potato virus Y (PVY) and potato virus X (PVX), and development of maize resistant to the maize streak virus (MSV). The inventory indicates that no virus resistant GM varieties have been commercially released in Africa.

In the Eastern and Central Africa region, maize is a major staple of the rural and urban poor. In the same region, potato has become a major highland cash crop and a food staple in some urban areas. Sweet potato is an important crop in the countries surrounding Lake Victoria (Burundi, the Democratic Republic of the Congo, Kenya, Rwanda, United Republic of Tanzania and Uganda). The FAOSTAT database indicates that the 2003 production in Africa for sweet potato was 10 787 127 tonnes, potatoes 12 530 119 tonnes and maize 43 522 313 tonnes. Indigenous to Africa and its offshore islands, MSV causes yield losses of up to 100 percent even in high potential agricultural zones. MSV is considered the most serious virus threat to maize production in Africa, while SPFMV is a serious constraint to sweet potato production. In the Near East region, Egypt is conducting field trials on virus-resistant GM varieties of potato, tomato, cucumber, melon, muskmelon, cantaloupe, squash and sugar cane. Tunisia has initiated virus-resistant GM research work on potato and *Vitis,* and The Islamic Republic of Iran has initiated this research work on sugar beet. Most virus-resistant GM varieties under field testing are of imported origin. However, one country in this region, Egypt, is carrying out field testing of locally developed virus resistant varieties, mainly cucurbits with resistance to ZYMV, and potato resistant to PVY and PLRV.

At least two virus-resistant transgenic crop varieties have been commercially released in the Asia region. These include virus-resistant tomato and green pepper varieties in China. In general, current research efforts to develop transgenic virus-resistant crop varieties in this region are mostly focused on *Solanaceae, Cucurbitaceae* and tropical fruit trees. Special attention is being given to development of virus-resistant papaya, as transgenic papaya varieties resistant to papaya ringspot virus (PRSV) now provide an additional strategy to restore papaya cultivation in areas where the virus has been most destructive (e.g. in Hawaii where major successes have been reported). Research has been reported in Malaysia, the Philippines and Thailand on the development of different ring spotvirus resistant transgenic varieties of papaya, some being combined with delayed fruit ripening.

Field trials are being conducted in China for transgenic varieties exhibiting resistance to: CMV in sweet pepper, CMV and TMV in chilli pepper, TMV in tobacco, PRSV in papaya, stripe virus in groundnut, PVY in potato, BYDV in wheat and turnip mosaic virus in Chinese cabbage. Laboratory stage research work on RDV in rice and WYMV in wheat is in progress in China. Thailand has field trials in progress for transgenic virus resistant varieties of tomato (TYLCV), papaya (PRSV) and pepper (CVbMV), and laboratory stage work on pepper (PepLCV), yard long bean (aphid-borne mosaic virus) and rice (ragged stunt virus). The Philippines has field trials underway for virus-resistant transgenic varieties of banana (BTV), and laboratory stage experimental work on transgenic papaya (PRSV). Indonesia has laboratory research underway on the development of virus-resistant peanut (peanut stripe virus), tobacco (TMV), sweet potato (SPFMV), chilli pepper, papaya and potato (PVX and PVY) varieties. Malaysia is conducting transgenic experiments on the development of virus-resistant rice (tungro virus), papaya (PRSV), pepper (CMV) and chilli pepper, while Bangladesh is undertaking research on papaya to develop transgenic cultivars resistant to papaya mosaic virus. India and Pakistan are currently working on the development of virus-resistant rice, cotton, as well as tomato (Pakistan only).

Of all the regions covered by the FAO-BioDeC, Latin America has the highest number of reported research activities regarding the development of transgenic virus-resistant crop varieties. The FAO-BioDeC indicates that nine countries in Latin America are conducting research and/or development on transgenic virusresistant crop varieties: Mexico (14 reported activities), Brazil (13), Cuba (four), Argentina (two), Colombia (two), Peru, Chile and Costa Rica (two each), and Venezuela (one). No virus-resistant transgenic crop variety has been reported to be commercially released in this region. In Brazil, field trials are underway for virus-resistant transgenic varieties of sugar cane (SCMV, yellow virus), potato (PVY, PLRV), papaya (PRSV), tobacco (TSWV, PVY), bean (bean golden mosaic virus), *Solanaceae* (isolation of virus resistance genes), and tomato (Gemini and Tospovirus), and laboratory stage experiments are underway on virus-resistant sugar cane. Mexico has virus-resistance trials underway on varieties of papaya (PRSV), potato (PVY and PVX), squash (PAMV, SMV2 and ZaMV), tobacco (TMV), zucchini (PMV, PAMV, SMV2 and ZaMV), melon (CMV), and tomato (CMV). In Cuba, field trials of virus-resistant varieties of papaya (PRSV), and experiments on potato (PLRV), citrus (tristeza virus), and tomato (Gemini virus) are underway. Laboratory stage experiments to develop virus-resistant transgenic varieties are ongoing in Chile for potato and melon, in Costa Rica for rice and maize, in Venezuela for coffee and in Peru for potato and sweet potato.

The large number of research activities reported in the FAO-BioDeC, regarding genetic engineering approaches to generate virus-resistant crop varieties indicates that the basic molecular techniques are well established to develop transgenecassette based approaches for control of most crop viruses. This basic molecular biology capacity is seemingly available in many of the countries covered by the database and the transgene cassettes can often be developed or obtained through collaboration with partner laboratories in other countries with more advanced research capacity. The rapid adoption of this technology within the context of broader agricultural R&D reflects the widespread occurrence of viruses across the regions and crop species, and the difficulty of controlling crop viruses by conventional (non-transgenic) means.

Development of Transgenic Crop Varieties Resistant to Bacterial Diseases

The FAO-BioDeC provides details on the use of genetic engineering for the development of bacterial-resistant transgenic crop varieties. The results suggest a low level of R&D activity in this area in the five regions. Just one transgenic variety with enhanced bacterial resistance is reported as being under field trial—a potato variety with wilt resistance in China.

However, laboratory stage research initiatives to develop bacterial resistant crop varieties are reported in six countries: in China (potato and wheat wilt resistance); in Thailand (bacterial wilt and other resistances in tomato); blight resistance in Basmati rice in Pakistan; leaf blight in rice, potato and cabbage in the Republic of Korea; and also bacterial-resistance work on jute in Bangladesh and banana in Venezuela. The lower level of activity on resistance to bacterial diseases compared to other diseases may be due to both a lower perception of the importance of bacterial diseases and the number of crops infected by them compared to the incidence of viral diseases, and to more readily available alternative technologies to combat bacterial diseases.

Development of Transgenic Crop Varieties Resistant to Fungal Diseases

Some of the most devastating and universal crop diseases are caused by fungal pathogens. For instance, the rust fungi are the most widespread and generally cause the largest crop losses per season. Crop losses can be considerable due to fungal pathogens. For example, the fungal agent of rice blast disease (*Magnaporthe grisea)* destroys 157 million tonnes of cultivated rice each year, enough rice to feed 60 million people worldwide. The negative effects of some fungal pathogens can be limited by the use of chemical fungicides.

Demand for fungicides amongst farmers is high, indicating that for many farmers there are few available alternatives. The world market for agricultural and non-crop fungicides amounted to an estimated US$6 billion at the end-user level in 1999. The United States, Western Europe and Japan together accounted for 75 percent of the total world market. Small grains constitute the largest market for fungicides worldwide. This sector accounted for an estimated 27 percent of the total world market in 1999, followed closely by tree and vine crops (24 percent), rice (16 percent), and vegetables and potatoes (10 percent). Other crop markets accounted for 17 percent of the world fungicide market, and non-crop markets accounted for 6 percent.

In many countries, fungicides as crop protection products are subject to strict legislative regulation and undergo a rigorous and expensive process of registration for public sale. While fungicides can provide a level of control, this chemical option is often limited for many farmers, particularly in developing countries, by high costs and lack of knowledge about application. In addition, the negative effects of fungicide applications on human health, with special reference to the labourers and the environment can be considerable. There is a need to find more environmentally benign alternatives to fungicides to control fungal diseases of crops. Genes can be identified that confer resistance to fungal pathogens. For instance, many genes have been found that provide resistance to specific races of each rust pathogen.

In many cases, resistance genes are available in the gene pool of cultivated plants and can be transferred to them by cross-breeding programmes. The incorporation of plant-derived resistance genes against fungal pathogens into susceptible varieties could allow development of resistant varieties which can deliver high yields in the absence of fungicide applications. Actual and potential access by farmers to traditional fungal control measures such as fungicides, and the absence of durable transgenic genetic resistance strategies, may explain why there are few reported efforts to develop transgenic varieties which are resistant to fungal pathogens. In relation to fungi, the current stage of successful research worldwide on identification/isolation of genes conferring durable resistance to fungal diseases of crops, is probably not encouraging developing countries with scarce research resources to embark on transgenic approaches to fungal resistance.

In the African region, only two initiatives for fungal resistant transgenic varieties are reported, a field trial of transgenic strawberry with phytoalexin synthesis genes (e.g. *Vst1*, *Vst2*) and laboratory work on transgenic maize for resistance to cob rot (*Stenocarpella maydis*), both in South Africa.. The three initiatives for development of fungal resistant transgenic varieties reported in Eastern Europe were all in Bosnia and Herzegovina where laboratory testing of transgenic potato for resistance to *Fusarium*, *Verticilium* and *Rhizoctonia* has been initiated. In the Asian region, there is a field trial underway in China for transgenic cotton with resistance to *Verticilium* and *Fusarium*. Other R&D initiatives reported include the involvement of India, Malaysia and Pakistan in research on sheath blight of rice, and of Indonesia on rice blast and leaf rust of coffee.

A few countries in Latin America, mainly Argentina, Brazil and Cuba, are carrying out a number of activities on transgenic resistance to fungi, particularly on tropical fruit trees, with some results already being tested in the field. In this region, most of the activities for transgenic fungal resistance are reported in Cuba, in particular involving field trials of transgenic potato for late blight resistance, and fungal-resistant sugar cane. Other field trials in the Latin America region for transgenic fungal resistance are reported for maize, sunflower and wheat in Argentina, and tobacco in Mexico. Other crops subject to transgenic R&D for fungal resistance in Cuba are banana, plantain, pineapple, tomato, papaya, citrus and rice. Other countries involved in transgenic fungal resistance research are Argentina on alfalfa, Brazil on rice, barley and cocoa, Chile on grape and apple, Colombia on tree tomato, Peru on potato for late blight resistance and Venezuela on sugar cane.

Development of Transgenic Crop Varieties

Resistance to Insect Pests

Insects consume a large share of food and fibre destined for humans (pre- and post-harvest). The worldwide economic damage caused by insect pests to agricultural and horticultural crops and to orchards stands at a hundred billion dollars annually. The strategies to limit the damage from insect pests fall into three categories:

1. Treatment of crops and their pests with chemical insecticides. Drawbacks of the use of chemicals are the development of resistance by the insects to the insecticides and secondly, the toxicity of many pesticides for the environment and human health.

2. Development of insect-resistant crops by hybridization or by transgenic modification. While transgenic approaches to the development of insect resistance are likely to be the fastest route, or the only means when pest resistance is not available in the genetic pool of the crop plant considered, they are more controversial, as with all genetic modifications of crop plants.

3. Introduction of natural enemies (usually alien species introductions) to further biological balance with limited damage. This approach is limited by the availability of natural enemies of the target pest and often risky as it is difficult to predict the outcome of the changes in the agro-ecological niche.

For all these strategies, it is important to gain good knowledge of the pest species that may cause damage at a certain place and time. Quick and reliable identification of the species and monitoring of their geographical distribution and life history is the basis of all effective policies to control insect pests. Worldwide, there are several thousands of insect pest species known to affect a range of crop species, each with their own characteristic damage, distribution and natural enemies. The knowledge and expertise in this scientific field is still limited and often insufficient to develop effective control strategies, especially in tropical or subtropical environments which characterise most developing countries.

Development of Transgenic Crop Varieties Resistant to Lepidopteran Pests

Among the insect pests, *Lepidoptera* represent a diverse and important group. The FAO-BioDeC indicates that most insect-resistant transgenic crop varieties under R&D are obtained for the control of *Lepidoptera,* predominantly using transgene cassettes including a toxin-producing gene from the soil bacterium *Bacillus thuringiensis* (Bt). In Asia, *Lepidoptera*-resistant cotton varieties developed using Bt transgenes have been commercially released in China, India and Indonesia and field trials are underway in Thailand. Other field trials underway in China involve local varieties of rice containing Bt transgenes, transgenic cotton expressing Cowpea Trypsin Inhibitor (CPTI), rice transformed for resistance to stem borer and yellow borer, maize for corn borer, cotton for bollworm, and poplar for gypsy moth. Bt tobacco is also in field trials in India and Bt maize in Indonesia.

Laboratory stage research work is ongoing in Asia on potato to generate resistance to tuber moth in India and Indonesia, diamond back moth (cabbage) in India, in addition to generation of resistance to *Lepidopteran* pests of cotton, rice and chickpea in Pakistan. Bt maize has been commercially released in the Philippines, research is underway in Indonesia on transgenic maize for corn borer, soybean for pod borer, rice for stem borer, *Cacao* for fruit borer, sugar cane for stemborer and oil palm for *Setothosea asigna*. Research is also ongoing on transgenic cotton for bollworm resistance in China and Thailand, jute for hairy caterpillar in Bangladesh, soybean in China and Chinese cabbage in the Republic of Korea .

In Latin America, transgenic maize resistant to *Lepidopteran* pests has been released in Argentina, and is in field trials in Brazil and Mexico, and undergoing laboratory testing in Cuba. Other field trials are underway on transgenic soybean in Argentina and Brazil, and cotton in Bolivia, Brazil and Mexico, potato in Mexico and Peru, sugar cane in Brazil and Cuba, tomato in Mexico, sweet potato in Cuba and sunflower in Argentina. Laboratory stage work on transgenic resistance to *Lepidiopteran* pests of rice, coffee and pineapple is ongoing in Cuba and on sugar cane in Brazil. In the Near East there are field trials of *Lepidoptera*-resistant transgenic varieties of maize and potato in Egypt and rice in the Islamic Republic of Iran , and laboratory work on cotton and maize in Egypt and the Islamic Republic of Iran, respectively. Only four activities to generate *Lepidoptera*-resistant transgenic varieties are reported for Africa. Both *Lepidoptera*-resistant transgenic maize and cotton are under commercial cultivation in South Africa, and laboratory work is underway in Kenya. Transgenic Bt cotton is reported to be under field trial in Zimbabwe. The rapid adoption of the use of Bt derived transgenes for development of pest-resistant crop varieties, suggests that this approach has generated a new and valuable option for control of some crop pests, which can be applied across a range of agro-environments, crop species and pests.

Development of transgenic crop varieties resistant to Coleopteran pests

Beetles and weevils are also important insect pests of crops. A wide range of beetles are of economic importance since they interfere with agricultural and forestry crops, timber products and stored products, etc. However, beetles do not transmit any diseases of humans or livestock. Due to the fact that there are many predators, herbivores and scavengers amongst the beetles, they play an important role in maintaining the ecological balance in natural systems.

Furthermore, many hostspecific species are used as biocontrol agents of insect pests and noxious weeds. As an example of *Coleopteran* pests, the African sweet potato weevils (*Cylas puncticollis* and Cylas *brunneus*) are major pests of sweet potato production in sub-Saharan Africa. Sweet potato crop losses due to weevils range from 20-100 percent with more severe losses reported in the dry season or during droughts. No genes conferring durable resistance to such weevils have yet been identified within the sweet potato gene pool. The case of sweet potato weevils in Africa is an interesting one regarding decisions on the most effective options for control of the weevils. There are options for the control of other weevils (e.g. *Cylas formicarius*) used in other regions of the world, which are based on the use of pheromones and bio-insecticides. However, control of African sweet potato weevils has not been successful in Africa and bio-insecticides are generally not available or affordable to African sweet potato farmers. In such a context, the generation of weevil resistant crop varieties using transgenic technology presents a new and potentially valuable option for weevil control.

Development of Transgenic Crop Varieties Resistant to Nematode Pests

Nematodes are distributed worldwide and live saprotrophically or as parasites of plants, animals and humans. While most nematodes in soil are actually beneficial, farmers are most concerned with nematodes that are pathogens of the roots, stems, leaves or seeds of plants. Plant parasitic nematodes are of great economic importance, they are responsible for over US$100 billion in annual crop losses worldwide.

They attack a wide variety of plant species, among them many staple crops, vegetables and ornamentals. The root-parasitizing nematodes of the genera *Meloidogyne* (root-knot nematodes), *Heterodera* and *Globodera* (cyst nematodes) are the most important pathogens. While substantial yield increases have been realised with the discovery and use of nematocides, considerable work remains to be carried out in many aspects of nematode biology and management. On average, only 0.2 percent of the value of crop loss due to nematode damage is invested into nematode research. Furthermore, nematology is a relatively young science and although many crops suffer losses due to nematodes, the number of nematologists working in agriculture is still largely insufficient to properly address this problem. Therefore, much remains to be learned to manage these serious pests.

Due to high crop losses, there is generally a need for the treatment of infested soil with chemical nematocides. Many synthetic chemical nematocides are very unspecific and possess a high general toxicity so that they have been, or will soon be, banned by many governments. Alternative efficient methods to control nematode infestations are rare. Therefore, the search for new nematode control methods is of great economic interest for developing countries. Strategies deploying crop genetic resistance to nematodes could generate new options for nematode control. The FAO-BioDeC contains no reports in all regions on the use of transgenic technology to develop nematode resistant crop varieties. This was surprising given the high frequency and ubiquity of nematode pest species and the wide range of potential hosting species, across all the regions surveyed. The lack of reports of R&D to develop nematode resistant transgenic crop varieties might be influenced by three factors:

1. the remaining existence on the market of some nematocides that maintain their effectiveness for a considerable number of growing seasons. Many of these nematocides are no longer patent-protected and are available at reasonable cost to some farmers;

2. the limited geographic mobility of the nematode species, which spreads from one field to another very slowly. In many cases, therefore, farmers tend to solve the nematode problem inexpensively by transferring cultivation to other nearby fields; and

3. the slow progress to date in identification of monogenes conferring efficacious resistance to nematodes.

Development of Transgenic Crop Varieties Resistant to Herbicides for Weed Control

Adequate systems for weed control are an essential component of all farming systems.

Depending on the crop and weeds present, uncontrolled weeds can reduce yields by over 50 percent, impair crop quality, contaminate the harvest with undesirable weed material, and increase the likelihood that the crop will be attacked by insects or diseases. Weeds compete for nutrients and light, especially during the early stages of crop growth, and for moisture in drought stressed areas, often causing severe yield losses. For example, in labour intensive, small farm operations on the Nigerian savannah, weed-related yield losses from 65 to 92 percent have been recorded.]

The parasitic flowering plant known as "witchweed" *(Striga spp.)* remains a major pest of staple crops in sub-Saharan Africa, despite a number of unsuccessful initiatives and considerable research input to find a means of control. The areas infested with parasitic broomrapes (*Orobanche* spp.) and witchweeds are vast and expanding. *Striga* is considered as the greatest single biotic constraint to food production in Africa, where the livelihood of 300 million people is adversely affected. In infested areas, yield losses associated with *Striga* damage are often significant, ranging from 40-100 percent. Crop yields could potentially be doubled if such weeds could be controlled. However, labour intensive weeding is largely ineffective against weeds like *Striga.*

Weeds can be controlled mechanically (by cultivation or hoeing), chemically (with herbicides) or agronomically (e.g. crop rotation). In addition to the undesirable effects of weeds on agricultural production, many weeds can also damage natural areas, alter ecosystem processes and facilitate displacement of native species. Herbicide tolerant varieties provide new options for the control of major weeds which are constraining agricultural production in the regions surveyed.

In relation to herbicide resistant transgenic varieties, the situation is rather different among regions and most of the R&D activity in this area seems to be concentrated in a small number of countries per region. To date, the database suggests that research attention has been focused largely on the development of transgenic crop varieties resistant to glyphosate (Roundup) and glufosinate ammonium (or phosphinothricin, commercialised under the name Basta). There are also reports of some instances of development of crops resistant to herbicides such as bromoxynil, imidazolinone, bialaphos and isoxazoles.

Glyphosate resistance has been most widely used in Latin America, where Roundup-ReadyTM (RR) transgenic soybean has been cultivated in Argentina and Uruguay. According to other reports, the same two countries have released two more herbicide-resistant soybean varieties, however, without specification of the kind of herbicide resistance. A glufosinate-resistant transgenic maize variety has also been released in Argentina. Brazil has recently accepted the RR soybean and is now cultivating it on a large scale. Thirty-four herbicide tolerant transgenic varieties are being field tested in Latin American countries, particularly in Argentina, Brazil and Mexico. In this regard, field trials have been performed for transgenic cotton, alfalfa, maize, sunflower, sugar beet and wheat in Argentina, maize, cotton, sugar cane and eucalyptus in Brazil, and cotton, maize and soybean in Mexico.

In addition to glyphosate, glufosinate resistant varieties are important in Latin America, where 14 field trials have been carried out in the region on transgenic soybean in Argentina, Brazil and Mexico, sugar cane in Brazil and Cuba, maize in Brazil and Mexico, wheat in Argentina and Mexico, sugarbeet in Argentina, rice in Brazil and potato in Cuba, which also has extensive laboratory input into rice, banana, plantain, coffee and pineapple. However, it must be noted that the number of laboratory research activities in the area of glufosinate resistance is much lower, and all such research activities are concentrated in Argentina (barley and sugar cane), Cuba, (rice, banana and plantain, coffee and pineapple) and Venezuela (sugar cane and mango). In Eastern Europe, field trials on glyphosate and glufosinate tolerant maize are ongoing in Serbia and Montenegro, whereas Bulgaria is commercially cultivating unspecified herbicide tolerant maize. In the Near East, the only research reported is on transgenic canola in the Islamic Republic of Iran. In Asia, Indonesia is conducting herbicide tolerance field tests on imported GM maize, cotton and soybean, and China on soybean. China is conducting research on herbicide tolerant rice, Pakistan on wheat and the Republic of Korea on cabbage, chinese cabbage and potato, though no details are given.

In Africa, the only GM herbicide tolerance research reported in the FAOBioDeC is taking place in South Africa, with field trials of 11 transgenic varieties resistant to different herbicides. The crops involved included canola, cotton, *Eucalyptus,* lucerne, maize, soybean, sugar cane and strawberry. In all regions, the higher number of field trials when compared to laboratory stage research in this area, suggests that most of the varieties undergoing field trials are likely to have been generated outside of the regions or result from crosses with local varieties. In addition, the relatively high number of laboratory research in this area compared to e.g bacteria resistance may also reflect the comparative technical simplicity for many species of generating herbicide tolerant crops using transgene cassettes obtained from research institutions or companies with more advanced research capacity. The limiting factors to R&D to generate herbicide tolerant transgenic crops are the number and type of herbicides and transgene resistance cassettes available, the ease with which the crop species or variety can be transformed, the registration requirements both for the herbicide and the transgenic crop variety, and the level of access of farmers and researchers to proprietary herbicides and transgene resistance cassettes. Yet, the early experiences with herbicide resistant maize to control *Striga* infestations in Africa suggest that novel models to disseminate herbicide tolerant transgenic crops could benefit even resource-poor farmers.

Transgenic Varieties Resistant or Tolerant to Abiotic Stresses

Abiotic stresses continue to limit crop productivity in every season and in every agro-ecosystem worldwide. Among the abiotic stresses affecting crop production in developing countries, drought and low soil fertility are the most significant. Plants vary tremendously in their ability to withstand abiotic stresses, both between species and within populations of a single species, but the nature of abiotic stress tolerance is not well characterised. Understanding the mechanisms of abiotic stress tolerance will have a significant impact on crop productivity. For instance, crop loss to drought in the tropics alone is thought to exceed 20 million tonnes of grain equivalent per year, or around 17 percent of well-watered production, reaching up to 60 percent in severely affected regions such as southern Africa in 1991-92.

The development of crop varieties (transgenic or non-transgenic) which can tolerate abiotic stresses would be of major benefit to agriculture in cropping regions where abiotic stresses are a chronic problem.No transgenic crop variety tolerant to abiotic stress has so far been reported to be released for cultivation in any of the regions covered by the database. However, 7 transgenic varieties exhibiting tolerance to a range of abiotic stresses underwent field testing in Bolivia (a frost tolerant potato variety), China (a cold tolerant tomato), Egypt (a salt tolerant wheat variety), India (moisture stress tolerant *Brassica* variety) and Thailand (salt tolerant and drought tolerant rice varieties). The number of research initiatives at the laboratory stage in this area totals 27.

Most of the R&D activities on development of abiotic stress tolerant crops are being carried out in six countries of the Asian region, namely Bangladesh, China, India, Indonesia, Pakistan and Thailand. China is relatively active in this area, and has reported preliminary successes with rice, maize and sorghum tolerant to high salt concentrations. Transgenic studies on salt resistant crop development are also being undertaken on rice in Bangladesh, Brazil, India and Pakistan, and on tobacco in Argentina. Transgenic research approaches to obtaining aluminiumresistant varieties are underway for wheat in Mexico and sugarbeet in China. Despite the effects of drought on crop production, very little transgenic research on drought resistance is reportedly being carried out in the five regions. Drought-tolerant rice is currently being field tested in Thailand. The rest of the activities in this area are mainly at the laboratory stage, with work on sugar cane (Indonesia), rice (China and Indonesia), and groundnut (South Africa).

Overall, the extent of R&D reportedly devoted to abiotic stress tolerance is insufficient when compared to the well known needs for abiotic stress tolerant varieties in the regions surveyed. For instance, vast areas of soils containing an excess of heavy metals are present in Brazil and Africa. Also, a steadily increasing acreage of agricultural land in Asia and elsewhere is becoming agriculturally sterile because of salinity brought about by poorly managed irrigation practices. The major limitation is the complexity of tolerance to abiotic stresses that is normally dependent on a number of physiological traits, each under multigenic control. The increased wealth of knowledge that is being acquired by means of genomics, mainly functional genomics, and other molecular biology studies, will certainly contribute to the development of tolerant genotypes.

Transgenic Varieties with Modified Quality Traits

The nutritional and economic value of many grain crops depends on the quantity and types of compounds (e.g. protein, starch and oil) that accumulate during seed development. Plant breeders have long strived to develop new crop varieties with enhanced quality trait profiles. A wide range of characteristics can fall under the topic of 'quality trait' and the use of this loose term in a survey can result in different categorisations by different respondents. For instance, the reports from Latin America indicate that there are field trials of varieties with enhanced quality traits underway on rice (Sucrose-phosphate synthase) and other trials for unspecified characteristics are underway on soybean in Argentina, and canola and flax in Mexico. Laboratory studies cover alfalfa (veterinary edible vaccines), *Paspalum dilatum* (foddering quality), *Triticale* (biomass production) in Argentina, eucalyptus (reduction of lignin content) and *Psychotria* spp. (improved alkaloid production) in Brazil, sugar cane (altered lignin content and high quality sugar) in Cuba, *Echinaceae*, *Psychotria* and *Tagetes* (alkaloid production) in Costa Rica, and improvement of flour quality and reduction of natural toxicants in potato in Peru.

Quality traits also exist in the floriculture industry. In this regard, a petunia variety with altered flower colour is now in commercial production in China. Micronutrient malnutrition, especially lack of iron, zinc and vitamin A, currently afflicts more than half the world's population. Called "micronutrients" because they are needed in only miniscule amounts, these substances enable the body to produce enzymes, hormones and other substances essential for proper growth and development. Tiny though the amounts are, the consequences of their absence are severe. Iodine, vitamin A and iron are most important in global public health terms. Their lack represents a major threat to the health and development of populations the world over, particularly to pre-school children and pregnant women in low-income countries. Enhancing the micronutrient (vitamin and mineral content) status of staple crops is considered to be one approach where crop biotechnology could generate crop varieties that could be used to strengthen food security and prevent malnutrition.

Protein Content and Amino-acid Profiles

Protein-energy malnutrition (PEM) is a most lethal form of malnutrition. Children are its most visible victims. Malnutrition contributes to at least half of the 10.9 million child deaths each year. PEM disorders result from a lack of protein and carbohydrates (energy). Cereals and root crops such as cassava form a substantial portion of developing country staple diets, thus improving their protein content and quality will go a long way to curbing the devastating effects of PEM. Seed proteins of many major crops often do not contain sufficient quantities of amino acids essential in the diet of humans and other monogastric animals. As cereals are usually deficient in lysine and tryptophan, and legumes are deficient in the sulphur amino acids methionine and cystine, a mixture of cereals and legumes is used to provide a balance of amino acids in the diet. Breeders at CIMMYT (Mexico) have managed with conventional breeding (over three decades) to develop quality protein maize (QPM) varieties which have enhanced levels of the two 'essential' amino acids, lysine and tryptophan, in the endosperm protein. These new varieties look and taste like normal maize but the nutritive value of their protein is nearly equivalent to cow's milk. Modifying the genes that encode seed proteins by genetic engineering is one additional option to address the problem of nutritional quality in some staple crop varieties. The FAO-BioDeC does not reveal any significant level of research on metabolic engineering of protein content nor quality in crop varieties in the vast majority of countries surveyed. The only research reported is on field trials for high lysine maize in China. Argentina is also reported to have conducted field trials for unspecified traits in maize and soybeans as well as wheat with high molecular weight glutein.

Vitamin Profiles

Vitamins are essential components of the human diet and dietary deficiencies in some vitamins can have tragic effects. For instance, vitamin A deficiency (VAD) is the leading cause of preventable blindness in children and raises the risk of disease and death from severe infections. In pregnant women, VAD causes night blindness and may increase the risk of maternal mortality. It is a public health problem in 118 countries, especially in Africa and South-East Asia, once again hitting hardest young children and pregnant women in lowincome countries. Crucial for maternal and child survival, supplying adequate vitamin A in high-risk areas can significantly reduce mortality.

The arsenal of nutritional approaches to combat VAD includes a combination of breastfeeding and vitamin A supplementation, coupled with enduring solutions, such as the promotion of vitamin A-rich diets and food fortification. Among these approaches, biotechnology research has contributed the additional option of increasing the vitamin A content of staple foods which are typically low in vitamin A content. The proof of concept of this approach has been the highly publicised high-beta carotene (pro-vitamin A) rice transgenics recently developed, and currently under dissemination to many NARS, under the name of 'golden rice' and 'golden mustard' in India. Yet, the FAO-BioDeC indicates that there is little R&D underway on enhancement of vitamin content in crop varieties in the regions surveyed.

Mineral Profiles

Most staple crops are not considered an important source of minerals in the human or domestic animal diet. Yet, because of the high level of consumption of staples, small increases in the mineral concentration of staples could have a significant effect on human nutrition and health. Iron deficiency anaemia afflicts an estimated billion people in developing countries, most of them women, reducing mental ability, creating severe complications at childbirth, and lowering physical capacity. Zinc deficiency, though less well understood, is also known to be widespread in the tropics and is a major threat to children's growth and health. The nutritional quality of staple crops (rice, cassava, wheat, maize and beans) in terms of mineral content can be improved by both conventional breeding and/or biotechnology. Currently the FAO-BioDeC does not report any biotechnology activity targeted towards the improvement of crop mineral profiles in developing countries.

Oil Composition

It seems from the countries covered by the FAO-BioDeC that the modification of plant oil and wax composition has received little attention. Among the instances of R&D in this area are ongoing field trials in Argentina on maize and soybean, and in Mexico on canola expressing high levels of lauric acid. In addition, there is ongoing laboratory work on oil palm with low saturated fatty acids in Indonesia and Malaysia and high lauric acid content in coconut in the Philippines. Furthermore, in Malaysia there is development of oil palm with special oils for the production of biodegradable plastics.

Plant Growth Traits

The genetic manipulation of the metabolic routes leading to plant hormone (e.g. ethylene) synthesis and degradation can induce modifications of plant organ maturation. Such control of maturation can allow the production of fruits showing resistance to postmaturation deterioration, resulting in the ability to be transported without refrigeration, extended shelf life and improved quality. For many developing countries with limited refrigeration capacity and transport infrastructures, this research avenue is extremely promising, insofar as it improves flexibility of transport of otherwise perishable products to distant markets andon poor routes.

Most of the research reported on alterations in plant growth has concentrated on delayed fruit ripening. In Latin America, Mexico reports a large number of field trials for altered growth in varieties of crops such as tomato (two approaches), chilli pepper, banana, melon, papaya and pineapple. In this area, Chile has a laboratory research initiative on stone fruits. In Asia, papaya varieties with altered growth traits are under laboratory study in China, Indonesia and the Philippines. In the latter, mango is also being researched, as is tobacco for delayed leaf senescence. Other growth phase changes receiving attention are short-stature and profuse tillering in basmati rice in Pakistan. In Eastern Europe, laboratory work in this area is ongoing in Bosnia and Herzegovina in potato for higher cytokinin levels.

'Stacking' Transgenes—Generating Crop Varieties with Multiple Transgenic Traits

The first generation of commercially available transgenic crop varieties has typically included cases where one transgene has been used to add an enhanced characteristic to a particular crop variety. Yet, there are a multitude of characteristics which are assessed by farmers to determine whether a variety is suitable or unsuitable. The production of crops with multiple genetically engineered traits may seem a logical step to follow and, as more locally adapted transgenic varieties become available and accepted on the market, these will become the source material into which novel genes will be incorporated. However, some problems (e.g. trait silencing due to similarities between transgene cassettes or epistatic interactions between transgenes) can arise if too many transgenes are incorporated into a single variety, and such effects may be cumulative over successive generations. To overcome such problems, novel approaches to simplify and improve the process of introducing multiple transgenes into crop varieties are under development.

Where data on the use of multiple transgenes to develop single crop varieties are present in the FAO-BioDeC, the most commonly used dual-transgene combination is herbicide resistance combined with insect resistance. In Latin America, Argentina has released transgenic maize with *Lepidoptera* resistance and glufosinate tolerance and there are 11 field trials reported in the region on GM crops. The absence of laboratory work suggests that this material is a technology import. Transgenic traits of *Lepidoptera* resistance with glyphosate tolerance are undergoing field trials in alfalfa, maize, soybean and sunflower in Argentina, and in cotton in Argentina, Brazil and Mexico, where maize with *Coloeptera, Lepidoptera* resistance and glyphosate tolerance is also being field tested. *Lepidoptera* resistance combined with glufosinate tolerance is undergoing field testing in maize in Argentina.

In South Africa, two field trials tested insect resistance combined with bromoxynil or glyphosate tolerance, and maize with potential insect resistance and glyphosate tolerance was field tested. Other traits that are combined with herbicide tolerance are fungal resistance with glufosinate tolerance in wheat in Argentina, and unspecified pathogen resistance with glyphosate tolerance in Mexico. In Argentina, *Coleoptera* and PVY resistant potato is undergoing field trials.

In Asia, two field trials are ongoing in China to test transgenic rice for combined blight and RDV resistance and combined bacterial wilt and PVY resistance in potato, as well as salt and herbicide tolerance in potato. In the Philippines, experiments to combine fungal, insect and bacterial resistance with salt tolerance in rice are underway. Papaya is receiving attention for combined ringspot virus resistance and delayed ripening in the Philippines, and similar work to combine virus resistance with extended shelf life is reported in Malaysia. Golden rice is a multi-trait transgenic plant since it was made by the transfer of three different transgenes for the synthesis of â-carotene taken from daffodil (*Narcissus pseudonarcissus*) and bacterium *Erwinia uredovora*. Metabolic pathway engineering is technically complex and challenging, and according to the FAO-BioDeC it has not yet been carried out in any crop within a developing country setting.

REFERENCES

Bebawi, F.F. & Farah, A.F. 1981. "Effect of parasitic and non-parasitic weed on sorghum". *Experimental Agriculture*, 17: 337–341.

Boyer, J.S. 1982. "Plant productivity and the environment". *Science*, 218: 443–448.

Ejeta, G., Butler, L.G. & Babiker, A.G.T. 1992. *New approaches to the control of Striga*. Striga research at Purdue University. Bulletin RB-991, Agricultural Experimental Research Station. West Lafayete, Indiana, Purdue University, USA. 27 pp.

Lagoke, S.T.O., Parkinson, V. & Agunbiade, R.M. 1991. "Parasitic weeds and control methods in Africa. *In* S.K. Kim, ed. *Combating Striga in Africa*. Proc. of the Int. Workshop. Organised by IITA, ICRISAT and IDRC, 22–24 August 1988, IITA, Ibadan, Nigeria. 3-14.

Oerke, E.C., Dehne, H.W., Schonbeck, F. & Weber, A. 1994. *Crop production and crop protection: Estimated losses in major food and cash crops*. Amsterdam, Elsevier.

Pennisi, E. 2001. "The push to pit genomics against fungal pathogens. *Science*, 292: 2273-2274.

Tabashnik, B.E. 1994. "Evolution of resistance to *Bacillus thuringiensis*. *Ann. Rev. Entomol.*, 39: 47–79.

Ye, X., Al-Babili, S., Klöti, A., Zhang, J., Lucca, P., Beyer, P. & Potrykus, I. 2000. "Engineering provitamin A (-carotene) biosynthetic pathway into (carotenoid-free) rice endosperm". *Science*, 287: 303–305.

Bibliography

Adato, M. and R. Meinzen-Dick. 2003. "Assessing the impact of agricultural research on poverty and livelihoods", *Quarterly Journal of International Agriculture* 42(2):149-167.

Agarwal, S., and Nath, A.K. 2001. "Characterisation of peach (*Prunus persica* L.) cultivars using isozymes as molecular markers". *Sci. Hort.* 90: 227-242.

Ahn, S. N., Bollich, C. N. and Tanksley, S. D., RFLP analysis of genomic regions associated with cooked-kernel elongation in rice. *Theor. Appl. Genet., 1992, 84, 825-828.*

Al-Abdulkador, Ahmed M. 1992. *Economic Feasibility of Greenhouse Vegetable Production in Oklahoma.* Oklahoma State University, Stillwater. M.S. Thesis. 139 p.

Allard, R. W., *Theor. Principles of Plant Breeding,* Wiley, New York, 1960.

Alston, J.M., Norton, G.W., & Pardey, P.G. (1995). *Science Under Scarcity: Principles and Practice for Agricultural Research Evaluation and Priority Setting,* Ithaca NY: Cornell University Press.

American Soybean Association Conservation Tillage Study (2001).

Anderson, Robert G. and Robert Hadad. 1999. "Nutrient analysis of organic fertilisers for greenhouse vegetable production". *HortScience.* June. p. 463. Abstract.

Anon. 1991. "Boron for brambles and blueberries". *Northland Berry News.* March. p. 6.

Baenziger PS, Russell WK, Graef GL, Campbell BT (2006) "Improving lives: 50 years of crop breeding, genetics, and cytology" (C-1). *Crop Sci* 46: 2230-2244

Baxevanis A., Ouellette F. 2001. *Bioinformatics: A practical Guide to the Analysis of Genes and Proteins.* John Willey & Sons, Inc. NY., USA. 518p.

Bebawi, F.F. & Farah, A.F. 1981. "Effect of parasitic and non-parasitic weed on sorghum". *Experimental Agriculture,* 17: 337-341.

Bhat, S.R. and Srinivasan. 2002 "Molecular and genetic analyses of transgenic plants: considerations and approaches". *Nature* 349:726.

Boyer, J.S. 1982. "Plant productivity and the environment". *Science,* 218: 443-448.

Brennan, J.P. and Bialowas, A. (2001), *Changes in Characteristics of NSW Wheat Varieties, 1965-1997.* Economic Research Report 8, NSW Agriculture, Wagga Wagga.

Brook, Rhonda J. 2002. "Pollen tracker" . *Farm Industry News.* mid-February. p. 30-32.

Brookes G, Barfoot P (2007): GM crops: The first ten years-global socio-economic and environmental impacts. *AgbioForum* 9:1-13.

Browicz, K., and Zohary, D. 1996. "The genus *Amygdalus* L. (*Rosaceae*): Species relationship, distribution and evolution under domestication". *Genet. Resour. Crop Evol.* 43: 229-247.

Byczynski, Lynn. 1993. "Greenhouse tomato considerations". *Growing for Market.* April. p. 1, 12.

Byerlee, D. and K. Fischer. 2001. "Accessing Modern Science: Policy and Institutional Options for Agricultural Biotechnology in Developing Countries," *IP Strategy Today,* No. 1-2001.

Byrne, D.H., and Littleton, T.G. 1989. "Characterization of isozyme variability in apricots". *J. Am. Soc. Hort. Sci.* 114: 674-678.

Carpenter, J.E. and L.P. Gianessi. "Agricultural Biotechnology: Updated Benefits Estimates, National Center for Food and Agricultural Policy: Washington, D.C., January 2001.

Chu, S. *et al.*, The transcriptonal program of sporulation in budding yeast. *Science*, 1998, 282, 699-705.

Colley, Michaela, and Matthew Dillon. 2004. "The next great challenge: Breeding seed for organic systems" . *Organic Farming Research Foundation Information Bulletin*. Winter. p. 1, 4, 5, 29.

Conservation Tillage and Plant Biotechnology (CTIC) (2002): *How New Technologies Can Improve the Environment by Reducing the Need to Plough* (http://www.ctic.purdue.edu/CTIC/Biotech. html).

Conway, G., and V. Ruttan. 1999. *The Doubly Green Revolution: Food for All in the 21st Century*. London: Penguin Books.

Crow, J.F. 1998. 90 "Years ago: the beginning of hybrid maize". *Genetics*. 148:923-928.

Daar, Sheila. 1989. "Trichoderma biofungicide registered in U.S". *The IPM Practitioner*. September. p. 19.

Davenport G., Ellis N., Ambrose M., Dicks J. 2004. "Using bioinformatics to analyse germplasm collections. *Euphytica* 137: 39-54.

Duvick, D.N. 1997. "Genetic rates of gain in hybrid maize yields during the past 40 years". *Maydica*, 22:187-196.

Dweikat, I., Zhang, W. and Ohm, H., Development of STS markers linked to hessian fly resistance gene H6 in wheat. *Theor. Appl. Genet., 2002, 105, 766-770.*

Edberg, Kevin. 1994. "Alternate row cropping of Boyne raspberries". *Northland Berry News*. September. p. 16.

Edey, Anna. 1998. "Solviva: How to Grow $500,000 on One Acre and Peace on Earth". Trailblazer Press, Martha's Vineyard, MA. 221 p.

Ejeta, G., Butler, L.G. & Babiker, A.G.T. 1992. *New approaches to the control of Striga*. Striga research at Purdue University. Bulletin RB-991, Agricultural Experimental Research Station. West Lafayete, Indiana, Purdue University, USA. 27 pp.

El-Kharbotly, A. *et al.*, Segregation analysis and RFLP mapping of the *R1* and *R3* alleles conferring race specific resistance to *Phytophthora infestans* in progenies of dihaploid potato parents. *Mol. Gen. Genet., 1994, 242, 749-754.*

Evenson, R.E and D. Gollin. 2003. "Assessing the impact of the green revolution, 1960 to 2000". *Science* 300 (May): 758-762.

Evenson, R.E. and Gollin, D. (eds) (2003a), *Crop Variety Improvement and Its Effect on Productivity*, CABI Publishing, Wallingford.

Fisher, D.G., Deahl, K.L., & Rainforth, M.V. (2002); Horizontal Resistance in *Solanum tuberosum* to Colorado Beetle (*Leptinotarsa decemlineata* Say). *Amer. J. of Potato Res.* 79: 281-293.

Flaim, John. 1999. "Growing in haybales. *Small Farm Today*. August. p. 58-60.

Furman, J. L., M. E. Porter, and S. Stern. 2002. "The determinants of national innovative capacity". *Research Policy* 31(6), 899-933.

Gagnon, Bernard and Sylvain Berrouard. 1994. "Effects of several organic fertilisers on growth of greenhouse tomato transplants". Canadian Journal of Plant Science. January. p. 167-168.

Gebhardt, K., and Goldbach H. 1988. "Establishment, graft union characteristics and growth of *Prunus* micrograft". *Physiol. Plant.* 72: 153-159.

Gibson G., Muse S. 2002. *A primer in genome science*. Sinauer Ass. Sunderland, USA, 347p.

Gilles, G. *et al.*, Single nuclotide polymorphic discrimination by an electronic dot blot assay on semiconductor microchips. *Nature Biotechnol., 1999, 17, 365-370.*

Goulart, Barbara. 1991. "Raspberry trellises: Pitfalls and possibilities". *Northland Berry News*. June. p. 14.

Grant, V. (1981) *Plant Speciation* (Columbia Univ. Press, New York), 2nd Ed.

Guebert, Alan. 2001. "Supreme Court blesses plant patents; bye-bye bin-run seed" . *The Land* (MN). December 21. p. 3.

Guner N, Wehner TC (2003) "Survey of U.S. land-grant universities for training of plant breeding students" . *Crop Sci* 43: 1938-1944

Gupta, P.K., Balyan, H.S., Sharma, P.C., and Ramesh, B. 1996. Microsatellites in plants: a new class of molecular markers. Curr. Sci. 70: 45-54.

Hack C., Kendall G. 2005. "Bioinformatics: current practice and future challenges for life science education". *Biochem. Molecular Biol. Educ.* 33: 82-85.

Hallauer AR (2007) "History, contribution, and future of quantitative genetics in plant breeding: lessons from maize" . *Crop Sci* 47: S4-S19

Harborne, J.B. 1978. "Plant tissue culture and its biotechnological application". *Phytochemistry*. 17:1691-1692.

Hassall & Associates (2002), *Evaluation of the Co-operative Research Centre for Molecular Plant Breeding: Economic Evaluation Component,* Report prepared for the Co-operative Research Centre for Molecular Plant Breeding, in association with EconSearch, Sydney.

Hesse, C.O. 1971. "Monoploid peaches, *Prunus persica* Batsch: Description and meiotic analysis". *J. Am. Soc. Hort. Sci.* 96: 326-330.

Himelrick, David G. 1994. "Raspberry root rot". *American Fruit Grower*. June. p. 16.

Hochmuth, George J. 1992. "Production and economics of rockwool tomatoes in Florida". p. 40-46. In: Proceedings of the 13th Annual Conference on Hydroponics. Hydroponic Society of America, Concord, CA.

Huang, J., C. Pray, and S. Rozelle. 2002. "Enhancing the crops to feed the poor". *Nature* 418 (August): 678-684.

Hubbell, B. J., Marra M. C., and Carlson G. A., 2000. "Estimating the Demand for a New Technology: Bt Cotton and Insecticide Policies. *Am. J. Agric. Econ.* 82, 118-132.

Information: Safety Assessment of Potatoes Resistant to the Colorado Potato Beetle. Health Canada. April 1997.

Ismael, Y., Bennett, R. and Morse, S.. (2001), "Farm level impact of Bt cotton in South Africa." *Biotechnology and Development Monitor,* No. 48, p. 15-19.

James C (2006): *Global Status of Transgenic Crops, Various Global Review Briefs from 1996 to 2006*. International Service for the Acquisition of Agri-Biotech Applications (ISAAA).

Jennings, D.L. 1988. *Raspberries and Blackberries: Their Breeding, Diseases and Growth.* Academic Press, New York, NY. p. 87, 88.

Jones, Stephen. 2004. "Breeding resistance to special interests" . *OFRF Information Bulletin*. Fall. p. 4-7.

Karakousis, A. and Langridge, P. (2003), "A high-throughput plant DNA extraction method for marker analysis", *Plant Molecular Biology Reporter,* 21 95a-95f, March 2003.

King, Tim. 2004. "Growing organic seed fits farm's rotation" . *The Land*. December 17. p. 9A-11A.

Kirsten, J and M. Grouse. 2003. "The Adoption and Impact of Agricultural Biotechnology n South Africa." in *Economic and Environmental Impacts of First Generation Biotechnologies* ed. Nicholas Kalaitzandonakes, Kluwer Academic Press/Plenum Publications.

Kishore, G. M. and C. Shewmaker. 1999. "Biotechnology: Enhancing human nutrition in developing and developed world". *PNAS* 93: 5968-5972.

Kleiner, Bill. 1993. "Understanding orange rust on black raspberry". *Northland Berry News*. June. p. 14.

Lazarus W, Selley R (2005): *Farm Machinery Economic Cost Estimates for 2005,* Univ Minnesota Extension Service.

Lindner, B. (2004), "Privatised provision of essential plant breeding infrastructure", *Australian Journal of Agricultural and Resource Economics* 48(2), 301-321.

Marra, M.C., P.G. Pardey, and J.M. Alston, eds. *The Payoffs to Agricultural Biotechnology: An Assessment of the Evidence.* Washington: International Food Policy Research Institute, 2002.

McCann, D.A and Batterham, P. 1993. "Australian genetics: a brief history". *Genetica.* 90:81-114.

Morgante M., Salamini F. 2003. "From plant genomics to breeding practice". *Curr. Opin. Biotech.* 14: 214-219.

Morris, M., Dreher, K., Ribaut, J.-M. and Khairallah, M. (2003), "Money matters (II): Costs of maize inbred line conversion schemes at CIMMYT using conventional and marker-assisted selection", *Molecular Breeding* 11, 235-247.

Oerke, E.C., Dehne, H.W., Schonbeck, F. & Weber, A. 1994. *Crop production and crop protection: Estimated losses in major food and cash crops.* Amsterdam, Elsevier.

Olsson, G. 1986. *Svalof 1886-1986- research and results in plant breeding.* Svalof AB Press, Svalof.

Pauls K.P. 1995. "Plant biotechnology for crop improvement". *Biotechnology Advances.* 13:673-693.

Pennisi, E. 2001. "The push to pit genomics against fungal pathogens. *Science,* 292: 2273-2274.

Pray, C.E., D. Ma, J. Huang and F. Qiao. "Impact of Bt Cotton in China." *World Development,* 29:5 (May 2001).

Pritts, Marvin P. 1991. "Mulch those newly planted raspberries". *Organic Farmer.* Summer. p. 38-39.

Richey, F.D. 1927. "Convergent improvement in selfed alfalfa". *Am. Nat.* 61:430-449.

Robinson, R.A. 1996. "Return to Resistance: Breeding Crops To Reduce Pesticide Dependency". *AgAccess,* Davis, California, and IDRC Books, Ottawa, Canada.

Rudd S. 2003. "Expressed sequence tags: alternative or complement to whole genome sequences?" *Trenda in Plant Sci.* 7: 321-329.

Rundgren, Gunnar. 2003. "EU organic seed regulation adapts to realityv. *The Organic Standard.* July. p. 16.

Salunkhe, D.K.; Kadam, S.S. (1998). *Handbook of Vegetable Science and Technology.* Marcel Dekker.

Schaefers, George A. 1988. "Studying aphid-resistant brambles". *Fruit Grower.* June. p. 8.

Sen S., Churchill G. 2001. "A statistical framework for quantitative trait mapping". *Genetics* 159: 371-387.

Shelton, A.M., J.Z. Zhao, and R.T. Roush. "Economic, Ecological, Food Safety, and Social Consequences of the Deployment of Bt Transgenic Plants." *Annual Review of Entomology,* 47:845-81, 2002.

Shuford JA, Steckelberg JM, Patel R (January 2005). "Effects of fresh garlic extract on Candida albicans biofilms". *Antimicrob. Agents Chemother.* 49 (1): 473. doi:10.1128/AAC.49.1.473.2005. PMID 15616341.

Shull, G.H. 1908. "The composition of a field of maize". *Am. Breeders' Assoc. Rep.* 4:296-301.

Silagy C, Neil A (1994). "Garlic as a lipid lowering agent-a meta-analysis". *J R Coll Physicians Lond* 28 (1): 39-45. PMID 8169881.

Simmonds, N.W. (1991); Genetics of Horizontal Resistance to Diseases of Crops. *Biol. Rev.,* 66: 189-241.

Smith T.F., Waterman M.S. 1981. "Identification of common molecular subsequences". *J. Molecular Biol.* 147: 195-197.

Staff. 2004. "Should there be a GMO tolerance for organic?" *The Non-GMO Source.* April. p. 1-2.

Stebbins, G. L. (1950) Variation and Evolution in Plants (Columbia Univ. Press, New York).

Steiner M, Lin RS (June 1998). "Changes in platelet function and susceptibility of lipoproteins to oxidation associated with administration of aged garlic extract". *J. Cardiovasc. Pharmacol.* 31 (6): 904-8.

Tabashnik, B.E. 1994. "Evolution of resistance to *Bacillus thuringiensis. Ann. Rev. Entomol.,* 39: 47-79.

Trigo et al. (2002): *Genetically Modified Crops in Argentina Agriculture: An Opened Story.* Libros del Zorzal Buenos Aires, Argentina.

Tsitrone, A; Rousset, F and David, P. 2001. "Heterosis, marker mutational processes and population inbreeding history". *Genetics.* 159:1845-1859.

United Nations Secretariat, Population Division of the Department of Economic and Social Affairs. 2004. *World Population Prospects: The 2004 Revision.* New York: United Nations. Available at http://esa.un.org/unpp.

Walsh B. 2001. "Quantitative genetics in the age of genomics". *Theor. Pop. Biology* 59: 175-184.

Wiffig, Hans. 1997. "Organic red raspberry production in the PNW". *Northland Berry News.* Vol. 11, No. 2

Young, N.D., Zamir, D., Ganal, M.W., and Tanksley, S.D. 1988. "Use of isogenic lines and simultaneous probing to identify DNA markers tightly linked to the Tm-2a gene in tomato". *Genetics* 120: 579-585.

Index